T2-CRY-072

A BRAIN FOR ALL SEASONS

Books by William H. Calvin

*Lingua ex Machina**
The Cerebral Code
How Brains Think
*Conversations with Neil's Brain***
How the Shaman Stole the Moon
The Ascent of Mind
The Cerebral Symphony
The River That Flows Uphill
The Throwing Madonna
*Inside the Brain***

*with Derek Bickerton
**with George A. Ojemann

A BRAIN FOR ALL SEASONS

Human Evolution and Abrupt Climate Change

WILLIAM H. CALVIN

The University of Chicago Press ▸ Chicago and London

William H. Calvin is a neurobiologist, Affiliate Professor of Psychiatry and Behavioral Sciences at the University of Washington in Seattle. His Ph.D. (University of Washington, 1966) was in physiology and biophysics and his major research interest is in the Darwinian brain circuitry for higher intellectual functions. Several decades ago, his interest in the ape-to-human enlargement of the brain led him to an active interest in the ice ages, both anthropological and geophysical aspects. The last third of this book is based on his cover story for *The Atlantic Monthly*, "The great climate flip-flop."

WilliamCalvin@alum.mit.edu
http://faculty.washington.edu/wcalvin

The University of Chicago Press, Chicago 60637
The University of Chicago Press, Ltd., London

Printed in the United States of America

11 10 09 08 07 06 05 04 03 02   1 2 3 4 5
ISBN: 0-226-09201-1 (cloth)

Corrections and web links for this book may be found at
http://WilliamCalvin.com/BrainForAllSeasons/.

Library of Congress Cataloging-in-Publication Data

Calvin, William H., 1939-
A brain for all seasons: human evolution and abrupt climate change / William H. Calvin.
p. cm.
Includes bibliographical references and index.
ISBN 0-226-09201-1 (cloth ; alk. paper)
1. Human beings—Effect of climate on. 2. Human evolution.
3. Paleoclimatology. 4. Brain–Evolution. I. Title.

GN281.4 .C293 2002
304.2′5—dc21

2001037602

∞ The paper used in this publication meets the minimum requirements of the American National Standards for Information Sciences - Permanence of Paper for Printed Library Materials, ANSI Z39.48-1992.

Contents

Directions, Distances, Temperatures:

A westerly wind comes *from* the west, moving eastward.

Ocean currents, however, are described by the direction that the current would carry a ship *toward.*

An east wind and an east current thus flow in opposite directions. Don't blame the scientists for this one (it's an old maritime tradition); our peculiarities can be found in the extensive **glossary** starting at page 301.

For those who have to mentally recompute when traveling outside the non-metric countries (Burma, Liberia, and the United States): 16 kilometers is 10 miles, and a rise of 5°C is the same warming as 9°F. In the rest of the world, water freezes at 0°C and boils at 100°C, a comfortable outdoor air temperature is 25°C (77°F), and normal body temperature is a mere 37.0°C.

To the memory of my late friends,

DONALD N. MICHAEL (1923-2000)

DONALD J. REIS (1931-2000)

PENN G. GOERTZEL (1947-2001)

PATRICK D. WALL (1925-2001)

Preamble

ONE OF THE MOST SHOCKING scientific realizations of all time has slowly been dawning on us: the earth's climate does great flip-flops every few thousand years, and with breathtaking speed. Many times in the lives of our ancestors, the climate abruptly cooled, just within several years. Worse, there was much less rainfall in many places, together with high winds and severe dust storms. Many forests, already doing poorly from the cool summers, dried up in the ensuing decade. Animal populations crashed - and likely early human populations as well. Lightning strikes surely ignited giant forest fires, denuding large areas even in the tropics, on a far greater scale than seen during an El Niño because of the unusual winds. Sometimes this was only the first step of a descent into a madhouse century of flickering climate.

Our ancestors lived through hundreds of such episodes - but each became a population bottleneck, one that eliminated most of their relatives. We are the improbable descendants of those who survived - and later thrived.

There was very little food after the fires. Once the grasses got started on the burnt landscape, however, the surviving grazing animals had a boom time, fueled by the vast expanses of grass that grew in the next few decades.

HAD THE COOLING taken a few centuries to happen, so that the forests could have gradually shifted, our ancestors would not have been treated so badly. The higher-elevation species would have slowly marched down the hillsides to occupy the valley floors, all without the succession that follows a fire. Each hominid generation could have made their living in the way their parents taught them, culturally adapting to the shifting milieu. But when the cooling and drought were abrupt, surviving the transition was a serious problem. It was one unlucky generation that suddenly had to improvise amidst crashing populations and burning ecosystems.

And improvising meant learning to eat grass and the like, because that's about the only thing that grows in the first years after a fire. Back before agriculture, that meant managing to eat animals that had turned the grass into muscle. Alas, you have to catch such animals first and, whether rabbit or antelope, they're fast and wary. Small or big, they're best tackled by cooperative groups - and since a rabbit's meat can't be shared by very many people, hunters would have tried for the bigger grazing animals.

This had an interesting corollary. Even if a single hunter killed a big grazing animal, it was too much to eat - meaning that it was best to give away most of the meat and count on reciprocity when someone else succeeded. Even chimpanzees do this if they kill a bush pig or small monkey - and the handouts aren't limited to those that took part in the chase.

Such a climate-induced downsizing temporarily exaggerated the importance of such traits as cooperation, hunting, and innovation. We might call the survivors the Phoenix Generation, after the big bird of myth that arose from the ashes, over and over again. Centuries later, with the return of other resources, the hominid population numbers would have recovered and the traits essential during the bottleneck would have slipped in importance.

But several thousand years later, after the stories about the hard times had disappeared from the word-of-mouth culture, it happened all over again: another generation got surprised by a downside episode of the boom-and-bust climate. And this generation had to conduct another search for how to eat grass indirectly. Fortunately, their ancestors had survived the same challenge and some of the genes for the relevant behavioral traits were still present, waiting to be tapped again.

For the ones with the right stuff, the temporary savanna even offered a window of opportunity for expansion, a brief version of the expansion opportunity that the mammals experienced after the dinosaur extinction. And so this latter-day Phoenix Generation promoted those genes a little bit more, another stroke on the pump.

The ice-core record of temperature suggests that this phoenix scenario recurred hundreds of times, that the Phoenix paleoclimate pump is the longest-running rags-to-riches play in humanity's history. Even if each individual window of opportunity only changed the inborn abilities for hunting or cooperation by a mere one percent, 200 repetitions of this same selection scenario would (just like compound interest) be potentially capable of explaining seven-fold differences between our inborn abilities and those of our closest relatives among the great apes.

YET *HOW* did such abrupt coolings happen on a worldwide scale? And can such population oscillations account for the enormous increase in altruistic and cooperative behaviors in humans, compared to our closest cousins among the apes? Might they have set the stage for the emergence of language? The structured thinking needed for planning ahead or logical trains of reasoning? The survival skills of being able to regularly eat large grazing animals? For our reflective consciousness? And why didn't other land animals experience the same boost, given that they must have been put through the same trials? Why just us? (Short answer: Though they suffered from the bust, they weren't tuned into the grasslands boom-time aspect.)

Such are the questions tackled here during a trip to hominid settings in Europe and Africa, followed by an over-the-pole flight that looks down on the probable origins of the abrupt climate changes: great whirlpools in the North Atlantic Ocean near Greenland. They flush the cooled surface waters down into the ocean depths, part of a giant conveyor belt that brings more warm surface water into the far north. This keeps Europe - and, surprisingly, much of the rest of the world as well - a lot warmer, much of the time. Except, of course, when the northerly whirlpools fail. There are likely multiple ways in which this climate collapse can be triggered. The best-understood one is via the greenhouse effect. Gradual warming, paradoxically, can trigger abrupt cooling.

Many climate changes are not gradual affairs, like turning up a thermostat or ramping up a dimmer switch. A gradual greenhouse warming over several centuries is not how things usually happen. As when tilting a table, there's a point when things start to slide off, and a tipping point when the table flips into a sideways mode. Abrupt (by which I mean the year-to-decade time frame) climate changes are more like a light switch that suddenly, at some pressure, flips into an alternative state. Just as when a power surge injures a fluorescent light tube and it starts flickering between bright and dim, so warming can cause air temperature to start abruptly flickering between warm and cool - and so produce a madhouse century.

Were a cold flip to happen in our now-crowded world, dependent on agricultural productivity and efficient supply lines, much of civilization would be ruined in a series of wars over the shrinking food supply. With death all around, life would become cheap. Millions of humans would survive but those left would reside in a series of small countries under despotic rule, all hating their neighbors because of recent atrocities during the downsizing. Recovery from such antagonistic gridlock would be very slow.

Surprisingly, these large fast climate changes may be easier to prevent than a greenhouse warming or an El Niño. Maybe. *Maybe* is the good news.

HUMAN EVOLUTION from an apelike ancestor started about 5-6 million years ago. This ancestor probably looked a lot like the modern bonobo and chimpanzee, with which we share this common ancestor. It probably had a pint-sized brain and only occasional upright posture. We are, in a real sense, the third chimpanzee species, the one that made a series of important innovations.

The first departure from this chimplike ancestor was probably some behavioral change - but behavior doesn't fossilize very well, and so the first change we can observe in retrospect was that of the knees and hips. They shifted toward our present form, well adapted to a lot of two-legged locomotion. Then, much later, when the ice ages began, toolmaking became common and the brain began to enlarge and reorganize. So the period of hominid evolution breaks neatly into two halves, each several million years long: the period of adaptation to upright posture (plus heavens knows what else), and the period of toolmaking and brain enlargement (plus language and planning).

I'm one of the many scientists who try to figure out what's behind an interesting correlation: What did the ice ages have to do with ratcheting up our ancestor's brain size? Our australopithecine ancestors, though they were walking upright, had an ape-sized brain about 2.5 million years ago. Ape brains probably hadn't changed much in size for the prior 10 million years. But when the ice ages began 2.5 million years ago, brain size started increasing - not particularly in the other mammalian species, but at least in our ancestors. About 120,000 years ago, in the warm period that preceded our most recent ice age, modern type *Homo sapiens* was probably walking around Africa with dark skin - and sporting a brain that was three times larger than before the first ice age chatters 2.5 million years ago.

Now, it's not obvious what ice, per se, has to do with brain size requirements. Our ancestors would simply have lived closer to the tropics, were it too cold elsewhere. And it's not that much colder in the tropics during an ice age (most of us would likely rate it more comfortable). Something about the ice ages probably stimulated the brain enlargement, but neither average temperature nor average ice coverage seem likely to be the stimulus.

Climate change is, of course, a standard theme of archaeology, all those abandoned towns and dried-up civilizations. Droughts and the glacial pace of the ice ages surely played some role in prehuman evolution, too, though it hasn't been obvious why it affected our ancestors so differently than the other great apes. The reason for our brain enlargement, I suspect, is that each ice age was accompanied, even in the tropics, by a series of whiplash climate changes. Each had an abrupt bust-and-boom episode - and that, not the ice, was probably what rewarded some of the brain variants of those apes that had become adapted to living in savannas.

When "climate change" is referred to in the press, it normally means greenhouse warming, which, it is predicted, will cause flooding, severe windstorms, and killer heat waves. But warming could also lead, paradoxically, to abrupt and drastic cooling ("Global warming's evil twin") - a catastrophe that could threaten the end of civilization. We could go back to ice-age temperatures within a decade - and judging from recent discoveries, an abrupt cooling could be triggered by our current global-warming trend. Europe's climate could become more like Siberia's. Because such a cooling and drying would occur too quickly for us to make readjustments in agricultural productivity and associated supply lines, it would be a potentially civilization-shattering affair, likely to cause a population crash far worse than those seen in the wars and plagues of history. What paleoclimate and oceanography researchers know of the mechanisms underlying such a climate "flip" suggests that global warming could start one in several different ways.

For a quarter century global-warming theorists have predicted that climate creep was going to occur and that we needed to prevent greenhouse gases from warming things up, thereby raising the sea level, destroying habitats, intensifying storms, and forcing agricultural rearrangements. Now we know that the most catastrophic result of global warming could be an abrupt cooling and drying.

My specialty is the time when man was changing into man. But, like a river that twists, evades, hesitates through slow miles, and then leaps violently down over a succession of cataracts, man can be called a crisis animal. Crisis is the most powerful element in his definition.

–LOREN EISELEY, *The Night Country*, 1971

You might think of the climate as a drunk: When left alone, it sits; when forced to move, it staggers.

– RICHARD B. ALLEY,
The Two-Mile Time Machine, 2000

Down House

The home of Charles Darwin
See where Darwin wrote Origin of Species
Walk Darwin's 'thinking path' and gardens
Explore his revolutionary theories in the interactive exhibition
Audio tour • Tea room • Shop • Toilets
Opening hours
1 April - 30 September Wednesday - Sunday: 10am - 6pm
1 - 31 October Wednesday - Sunday: 10am - 6pm or dusk if earlier
1 November - 31 January Wednesday - Sunday: 10am - 4pm
1 March - 31 March Wednesday - Sunday: 10am - 4pm
Closed: February, Christmas Eve, Christmas Day & Boxing Day
To prevent overcrowding, timed visits operate.
Tickets are available from the house entrance.
For your safety
Regional Office Tel: 0171 973 3479

To: Human Evolution E-Seminar
From: William H. Calvin
Location: 51.4°N 0.1°E
Darwin's home (Downe, England)

Subject: **Catastrophic gradualism**

You signed up for this, I trust, because you read my preamble, all about seeing human evolution in the context of a bust-then-boom climate episode, with scattered groups surviving the fiery population crash. And like a phoenix arising from the ashes, going on to great things (well, at least, *us*) after a sufficient number of repeat performances, just one concentration and expansion episode after another, pumping us up.

I'm starting this little tour at Darwin's home, sitting at a park bench under a magnificent oak tree that dates back to Darwin's time here. Five years after returning from the voyage of the *Beagle*, Charles Darwin and his young family moved from central London to a pleasant country home about 16 miles to the southeast, near the village of Downe. He lived here forty years until his death in 1882. No more voyages around the world, not even trips to the Continent, but Darwin had correspondents everywhere, and sometimes they showed up at his front door.

And it was here at Down House that he raised pigeons, studied earthworms, and dissected barnacles. Here he sat, pen in hand, and wrote out his books that provided so much of our modern understanding of how nature came to be the way it is.

But only ten years ago, a scientific pilgrimage to Darwin's country home was remarkably difficult, unless you got directions from someone who had been here before. Only the most detailed guidebooks had a mention of Down House, and then only in the fine print. Get off the train from London at

Bromley South or Orpington, and the taxi driver, upon learning your destination, would knowingly suggest that there were much finer country homes to visit than Down House – clearly not understanding that it was Charles Darwin that made Down House so important, not its gardens.

If I were to give an award for the single best idea anyone has ever had, I'd give it to Darwin, ahead of Newton and Einstein and everyone else. In a single stroke, the idea of evolution by natural selection unifies the realm of life, meaning, and purpose with the realm of space and time, cause and effect, mechanism and physical law.
– DANIEL C. DENNETT,
Darwin's Dangerous Idea, 1995

Arrive, pay off your I-told-you-so taxi driver, and you'd find a low budget operation financed over the decades by the London surgeons, with only several rooms restored to what they were like in Darwin's day, back before the place had been turned into a girls boarding school in the early 20th century. There were a few rooms filled with old-fashioned museum cases laden with a dozen coats of paint, but most of the house was in sad need of repairs and unsuitable for visitors. And this for one of the great scientists of all time, not just one of England's greats.

Still, it was enormously inspiring to anyone who understood the intellectual triumph of Charles Darwin, this chance to see where he had thought it all through – his study with his microscope, his chair by the living room fireplace, and his "sand walk" out back, where he went for three walks a day to digest his thoughts. Often, one supposes, Darwin sloughed through the fine English rain, likely blowing in from the west after forming above the warm Gulf Stream.

Most people who think a little about evolution are wedded to the basic idea of gradual improvements in efficiency – and not much concerned with the *origins* of what was later improved (it was just "mutations"). Yet it was Darwin himself (a point omitted from even the modernized science exhibits at Down House) who first cautioned readers about getting fixated on efficiency, and who - at the same time - offered a route for invention. He noted that changes in function could be "so important," that an anatomical structure improved for one function could, in passing, serve some other function that utilized the same anatomical feature. (Darwin's example was the fish's swim bladder serving as a primitive lung.) Novelties come from those nascent secondary uses, not from a bolt out of the blue, as a cosmic ray mutates a gene.

IF YOU HAVEN'T SEEN DOWN HOUSE since the reopening in 1998, there's a lot more to see, thanks to much fund-raising by the British Museum. It is currently operated by English Heritage, which provides audio wands to guide you through the rooms.

Next to Darwin's study, there's his billiard room, where cause and effect operated on a simpler, more direct, level than it does in biology.

Across the hall is the large dining room with its bay windows; it was also the "justice room" where Darwin served as a magistrate on occasion.

The now-rebuilt stairway to the upstairs leads you to a series of former bedrooms, filled with modern exhibits about Darwin's science.

Darwin traveled into London for scientific meetings, but mostly he kept up an enormous correspondence. His was something like the modern "home office" style of working, that computers and communications are making possible even for scientists without inherited wealth. Darwin's life shows you another style of doing science, one without classes to teach or students to supervise, without grant applications to write, one where piecing together the big story operated alongside the careful dissection of barnacles, digesting it all on yet another loop around the Sand Walk, carrying a great stick which he struck loudly against the ground, making a rhythmical click as he walked along with a swinging gait.

The Sand Walk has what Darwin referred to as a "light side" with a long view of the rolling countryside and a "dark side" loop through the woods he had planted. It's when making your own third loop around the Sand Walk (now pebble covered in the familiar English Heritage style, though there are still some flints to be found) that you find yourself wanting to tell Charles Darwin about all that has happened in the last 130 years, about how he was right about Africa being the place where humans happened. Then you scale back your plans to something more suitable for the time it takes to make several more loops. I decide on abrupt climate change, since it shows how you can have catastrophic gradualism.

EXPLAINING THINGS VIA CATASTROPHES was seen then (as now) as a form of style without substance. It was simply too reminiscent of miracles. Gradual explanations were to be preferred, if they could be found. Jerks were to be avoided.

A nice algorithmic turning of the crank was, in comparison, a thing of beauty – and Darwin found a wonderful crank via his inheritance principle, where the more successful of the current generation were the ones who generated more of the minor new variants which future generations would test against their environment. Variations on a *successful* theme was the name of Darwin's game.

Darwin saw that the climate had changed many times – he immediately offered some geological details to support Louis Agassiz's 1837 notion of an ice age - and he assumed that animals and plants had to change too, to keep up with the times. The variants more in tune with the new environment would reproduce better, in turn spawning yet more variants around *their* gene type (many variants, of course, are worse than their parents, but they don't reproduce very well, what with high childhood mortality). So adult body characteristics could track the climate, thanks to some novelties proving to be heritable.

Efficiency improvements do, of course, result in the long run in a "lean mean machine," where many features not used for a long time are stripped out as excess fat. Until recent years, economists loved this view of things, with all its improving efficiency - until it became so apparent that it didn't explain an innovation, only its subsequent improvement. And in an economy dominated by market capture, where the first to market may overshadow a better late arrival, innovation is becoming much more important than efficiency. It's the "survival of the fastest."

THE NATURAL ASSUMPTION, surely valid in some cases, is that climate will change slowly enough for little improvements to track climate over the generations - say, more and more upright posture as the blister-like uplift of the East African highlands helped convert forests into open woodlands and then savannas. What I'd want to tell Darwin is that, just a decade ago, the ice cores revealed that there have also been very abrupt climate changes every few thousand years (on average; most are somewhat irregular exaggerations of an otherwise minor 1,500-year climate rhythm).

These jumps are superimposed on the better-known gradual trends arising from variation in the earth's orbit. They are so large and so quick that a single generation gets caught, forced to innovate on the spot - innovate behaviorally, that is, since there is no time for anything in the gradual adaptations line.

And this provides a way around the lean-mean-machine implications of traditional Darwinism. Continuing to carry around a lot of useful-in-a-pinch abilities is a good thing when, about once in every hundred human generations, the climate goes mad for a while. The variants that became lean mean machines didn't survive very well in the crunch.

Climate catastrophes are often mixed up with evolutionary jumps (imagined macromutations and the like). But when the climate catastrophes repeat so often, then a little one-percent change each time can jack us up, producing major changes in body and behavior in only a million years or so. Darwin, I like to think, would have been intrigued by this "catastrophic gradualism" insight.

Neither **glossary** items (starting at page 301) nor the **endnotes** (starting at page 317) are denoted in the text, to avoid superscript clutter. Consult them early and often.

The first turn in Darwin's Sand Walk,
as the "light side" turns into the "dark side."

From the Greeks to the nineteenth century there was a great controversy over the question whether changes in the world are due to chance or necessity. It was Darwin who found a brilliant solution to this old conundrum: they are due to both. In the production of variation chance dominates, while selection itself operates largely by necessity. Yet Darwin's choice of the term "selection" was unfortunate, because it suggests that there is some agent in nature who deliberately selects. Actually the "selected" individuals are simply those who remain alive after all the less well adapted or less fortunate individuals have been removed from the population.

– ERNST MAYR, *This is Biology*, 1997

To: Human Evolution E-Seminar
From: William H. Calvin
Location: 51.47794°N 0.29089°W 11m ASL
Evolution House, Kew Gardens

Subject: **The Darwinian Quality Bootstrap**

Back in London, I've been wandering around the Royal Botanical Gardens at Kew and its new Evolution House. It's a good place to think about all of those things that I didn't have time to tell my imaginary companion on the Sand Walk.

I'll try to make clear where the unresolved questions about human origins still lie. I've heard lots of them discussed at the various paleoanthropology, archaeology, primatology, and linguistics meetings that I've attended in the last several years. In the course of this e-seminar, I'll see if I can clarify this boundary between the known and the dimly seen.

WHAT, SOMEONE ASKED BY E-MAIL, did Darwin really discover? It probably isn't what you always thought.

It wasn't evolution per se. There had been an active public discussion of evolution since before Darwin was born (his grandfather Erasmus even wrote poems on the subject).

It wasn't adaptations to fit the environment, as the religious philosophers had already seized on that idea as suggesting design from on high.

Nor was it "survival of the fittest." That idea had been floated by Empedocles 2,500 years ago in ancient Greece, long before Herbert Spencer, in the wake of Darwin, invented the phrase we now use.

It certainly wasn't the basic biological and geological facts that Darwin discovered, although during his voyage around the world, and after discovering natural selection, Darwin did add quite a bit in the factual line.

What Darwin contributed was an idea, a way of making various disconnected pieces of the overall puzzle fit together, something like trying to solve a jigsaw puzzle without a picture for a model. He imagined the picture.

You do not recognize answers to questions that have never arisen.
– ALAN WALKER & PAT SHIPMAN, 1996

It wasn't, however, the idea of descent from a common ancestor. Diderot, Lamarck, and Erasmus Darwin had all speculated on that subject two generations earlier. And there were trees of descent around to serve as examples, given how by 1816 the linguists were claiming that most European languages had descended from the same Indo-European root language.

By 1837 Darwin had concluded that nature was always in the process of *becoming* something else, though again there had been other attempts like Lamarck's along this line. Darwin just looked at the biological facts in a different way than his predecessors and contemporaries, not forcing them to fit the usual stories about how things had come about. Fitting facts to an idea is a primary way in which progress is made in science, but a fit in one aspect has often blinded scientists to more overarching explanations.

But even that wasn't his main contribution. Charles Darwin had an idea that supplied a mechanism, something to turn the crank that transformed one thing into another.

BASICALLY, CHARLES DARWIN (in 1838 and, independently, Alfred Russel Wallace in 1858) had a good idea about the *process* of evolution, how one thing could turn into another without an intelligent designer supervising. Out of all the variation thrown up with each generation (even children of the same two parents can be quite unlike one another), some variants fit the present environment better. And so, in conditions where only a few offspring manage to reach adulthood (both

Wallace and Darwin got that insight from Malthus and his emphasis on biological overproduction), there is a tendency for the environment to affect which variants get their genes into the next generation.

Many are called, few are chosen by the hidden hand of what Darwin labeled "natural selection." The name comes from the contrast to animal breeding, so-called "artificial selection." It is, as Ernst Mayr noted, an unfortunate term, as it suggests an agent doing the natural selecting.

As Thomas Huxley said, when reading Darwin's book manuscript before its publication in 1859, "How stupid not to have thought of it before." Two and a half millennia of very smart philosophers trying to solve the problem, and then the answer turns out to be so simple. Like the Necker cube and similar perceptual phenomena, there are often several ways to look at the same facts, just as there are two equally valid roots of a quadratic equation, both of which give satisfaction. And "seeing" the alternative form can be difficult when your culture guides you to see the usual explanation. But the alternate form may lead you to a more coherent solution, one that also explains a much bigger jigsaw puzzle.

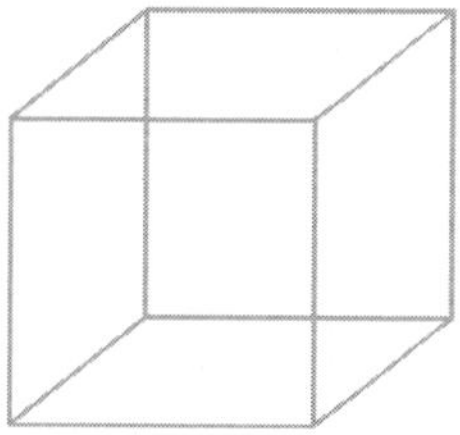

A few years later, Darwin realized that he needed to add an "inheritance principle," to emphasize that the variations of the next generation were preferentially done from the more successful of the current generation (the individuals better suited to surviving the environment or finding mates). This means, of course, that the new variations were not just at random, but were centered around the currently-successful model. In other words, they were little jumps from a mobile starting place, variations on a theme, not big jumps where the starting place becomes irrelevant because the jump carries so far. (Warning: Except for the pros, half of the people who write about evolution, whether pro or con, may be confused about this important short-distance randomness aspect.)

Many variations, of course, are not as good as the parents – nature appears not to worry about this waste, to our distress – but a few variants are even better than their parents. And so, with passing generations, there is a chance for drift to occur towards the better solutions to environmental and mate-finding challenges. Perfection you don't get, but occasionally you do get something that, locally, could be called "progress" – that ill-defined something that makes us so impressed by the Darwinian process. Nature can be seen to pull itself up by its own bootstraps, amidst a huge waste in variations that go nowhere.

YOU CAN SUMMARIZE Darwin's bootstrapping process in various ways, from our modern perspective. A century ago, Alfred Russel Wallace emphasized variation, selection, and inheritance. It reminds me of a three-legged stool: evolution takes all of them to stand up.

But there are some hidden biological assumptions in that three-part summary and, when trying to make the list a little more abstract to encompass non-biological possibilities for a Darwinian process, I wound up listing six ingredients that are essential (in the sense that if you're missing any one of them, you're not likely to see much progress):

1. There's a pattern of some sort (a string of DNA bases called a gene is the most familiar such pattern, though a cultural meme – ideas, tunes – may also do nicely).
2. Copies can be made of this pattern (indeed the minimal pattern that can be semi-faithfully copied tends to define the pattern of interest).
3. Variations occur, typically from copying errors or super-positions, more rarely from a point mutation in an original pattern.
4. A population of one variant competes with a population of another variant for occupation of a space (bluegrass competing against crabgrass for space in my backyard is an example of a copying competition).

5. There is a multifaceted environment that makes one pattern's population able to occupy a higher fraction of the space than the other (for grass, it's how often you water it, trim it, fertilize it, freeze it, and walk on it). This is the "natural selection" aspect for which Darwin named his theory, but it's only one of six essential ingredients.
6. And finally, the next round of variations is centered on the patterns that proved somewhat more successful in the prior copying competition.

Try leaving one of these out, and your quality improvement lasts only for the current generation - or it wanders aimlessly, only weakly directed by natural selection.

Many processes loosely called "Darwinian" have only a few of these essentials, as in the selective survival of some neural connections in the brain during development (a third of cortical connections are edited out during childhood). Yes, there is natural selection producing a useful pattern - but there are no copies, no populations competing, and there is no inheritance principle to promote "progress" over the generations. Half a loaf is better than none, but this is one of these committees that doesn't "get up and fly" unless all the members are present.

And it flies even faster with a few optional members. There are some things that, while they aren't essential in the same way, affect the rate at which evolutionary change can occur. There are at least five things that speed up evolution.

First is *speciation*, where a population becomes resistant to successful breeding with its parent population and thus preserves its new adaptations from being diluted by unimproved immigrants. The crank now has a ratchet.

Then there is *sex* (systematic means of creating variety by shuffling and recombination - don't leave variations to chance!).

Splitting a population up into *islands* (that temporarily promote inbreeding and limit competition from outsiders) can do wonders.

Another prominent speedup is when you have *empty niches* to refill (where competition is temporarily suspended and the resources so rich that even oddities get a chance to grow up and reproduce).

Climate fluctuations, whatever they may do via culling, also promote island formation and empty niches quite vigorously on occasion, and so may temporarily speed up the pace of evolution.

Some optional elements slow down evolution: "grooves" develop, ruts from which variations cannot effectively escape without causing fatal errors in development. And the milder variations simply backslide, so the species average doesn't drift much. Similar *stabilization* is perhaps what has happened with "living fossil" species that remain largely unchanged for extremely long periods.

You'll notice that I didn't even mention changes in the rate of mutations. Since sex and gene shuffling were invented, mutation rate may have fallen pretty far down the list of important factors controlling the pace of evolution, even though mutations are the usual beginner's example. Species shifts more often involve changes in the relative *proportion* of existing gene versions (gene frequencies). It's the committee's composition that counts; sometimes all it takes is removing one member to break a deadlock or open up new paths.

To: Human Evolution E-Seminar
From: William H. Calvin
Location: 51.3°N 1.4°E -40m ASL
Down among the fossils

Subject: **All of those chimp-human differences**

Hurrying through the fossils at eighty miles per hour, I can't see a thing except for some little lights that go by every two seconds, serving to reassure you that the train hasn't stopped somewhere beneath the ocean. A train usually jerks around enough to reassure the passengers that it hasn't stopped, but this undersea railroad is pretty smooth. I'm en route to a meeting in Paris, so you get my third installment for our virtual seminar from a tunnel ("the chunnel") that is 40 meters beneath the English Channel connecting Folkestone, England, with Calais, France.

Thinking back to the Sand Walk, I now realize that we often make mistakes in science by making logical but erroneous extrapolations. For example, "It's just a drop in the ocean" doesn't always scale up. We assume that oceans mix and dilute anything added, despite the evidence that oceans stay poorly mixed (and thus stay cold in the depths) because of dynamic processes that circulate in "streams" more quickly than diffusion can be effective. This is of some importance; were the depths to start warming up, CO_2 would come bubbling out just like from an uncapped bottle of seltzer. It would, of course, add to the greenhouse effect.

We don't scale up time very well either, which is why Darwin and the geologists of the nineteenth century had such a problem convincing nonscientists about the time scale of evolution. And they worked so hard in selling gradualism over eons that today, when we first begin to think about evolutionary change, we often assume that it is just like when you slide a cardboard box down a loading ramp. It doesn't accelerate very much by the time it reaches the bottom; it's not too different from pushing it across a horizontal floor.

But every beginning physics student soon learns how different fast dynamics is from slow statics. Process becomes important and oscillations may occur, as when a pendulum converts kinetic and potential energy back and forth. In nonlinear systems, dynamics is further complicated because there are often modes of operation, with change-of-state transitions between them that require a story of their own. The latent heats of the solid-liquid-vapor transitions are the best known – and appearances can fool you. A cloud layer on the lee side of a mountain may look perfectly static, but it is really forming by condensation at the bottom (which heats the water droplets) and evaporation at the top (which cools them), with a constant flow of water up through the cloud. The cloud is just an emergent property, its thickness a consequence of those change-of-state transitions.

For a century, we tried to pretend that evolutionary transitions were just like that box slowly sliding down a ramp, where a slight imbalance of forces operated slowly. But change-of-state transitions may become far more important than the longer-lasting states. For example, animals must survive and reproduce during a chaotic climate transition, when all of their hard-earned efficiency adaptations to a particular environment have just become worthless and a new regime hasn't yet been established.

OOPS, WE'VE POPPED OUT into France in a mere twenty minutes. Farther up the French coast is one of the German V-2 rocket-launching sites, made into a museum explaining how Hitler planned to invade Britain. Apropos of whether scientists are seen as saviors or serpents, I am reminded of Tom Lehrer's famous satire on scientific responsibility, "Once the rockets are up / Who cares where they come down? / That's not my department / Says Wernher von Braun."

This part of France looks suspiciously like Kansas – I confess, I graduated from a Kansas high school, and learned my evolution later. Hereabouts, the farm buildings have nice red tile roofs. Improbably, there are two-story-high earthen berms along the railroad's right-of-way wherever there is a nearby barn. These artificial hills deflect noise upward, perhaps keeping the cows from being disturbed by the noise of the Eurostar trains.

I'll get to ancient climate presently, but first I thought that I'd do a little stage-setting for this virtual seminar, a shopping list of *desiderata*. Since I don't want you to get *too* focused on higher intellectual functions (skimming the cream is, in general, a bad habit), this will be a short list of all the major behavioral changes since our last common ancestor with the chimps and bonobos, 5-6 million years ago.

Mine are all major improvements that need an evolutionary explanation for how we got beyond a great-ape level of abilities in so short a time. They are things like extensive *altruism* – not just the sharing with relatives that genes-in-common can help explain, but the sharing with (and taking risks to come to the aid of) strangers, where the giver gets paid back (if at all) by some third person. Also, things that develop trust among individuals in large groups who don't all know one another – a prime example is the worldwide scientific community.

SHARING AND COOPERATION may be good for the species in the long run, but evolution has no known foresight mechanism. This means that every little increment has to pay its own way with immediate advantages. It's difficult even to explain an in-between stage, reciprocal altruism – that's where you're doing favors for unrelated friends – because the entry-level stages of it are so easily swamped by freeloaders who receive without giving back. It's leaky, like a tire that sinks after a while.

Yet some reciprocal altruism exists, in chimps and bonobos, and our ancestors developed altruism more generally into disaster relief, welfare, and peacekeeping forces. How was such altruism bootstrapped up, through a series of stages with intermediate payoffs?

Similar questions apply to other beyond-the-apes abilities that we have. Yet, compared to stones and bones, such issues can be confusingly abstract, and so we tend to focus on "hard evidence" like the hip-and-knee rearrangements needed for upright posture. Or the big increase in brain size. We can, however, infer a lot of behavioral changes, just by comparing our fully-human abilities with the versions seen in our closest cousins.

So here's my most recent attempt at a hominid bootstrap list, for what happened along with upright stance and enlarged brains. In my opinion, the big beyond-the-chimps improvements that need step-by-step bootstrap explanations covering the last five million years are:

- *altruism* (beyond chimp-level reciprocal altruism),
- *accurate throwing* (not just flinging, which many chimps do, but practicing to hit smaller and smaller targets),
- *extensive tool making* (especially tools with which to make other tools),
- *protolanguage* (real words used in short combinations, such as the language of two-year-olds),
- *structured language* (long sentences with recursive embedding of phrases and clauses, likely a different evolutionary problem),
- *planning* for uncertain futures (not just the seasons) and their associated *agendas*,
- *logical trains of inference* that allow us to connect remote causes with present effects (and a propensity to guess at them, useful both for doing science and for fooling yourself),

- *ethics* (which may require an ability to estimate the consequences of a proposed course of action, and judge it from another's standpoint),
- *concealed ovulation* (the disappearance of obvious "in heat" periods tended to force males into prolonged sharing with a female and her previous offspring, just to be around at the right time),
- *games* with made-up rules (hopscotch, not just play) and dance,
- our fascination with discovering hidden patterns, seen in *music* (not just rhythm but four-part harmony), art and abstractions, crossword puzzles, and doing science, and
- our extensive offline *creativity* (an ability to speculate, to shape up quality by bootstrapping from rude beginnings, yet without acting in the real world).

You'll notice that I didn't use the C word here, though I did describe most of the beyond-the-apes uses of consciousness along the way.

There are many more differences, of course, but I'm trying here for ones with a big order-of-magnitude improvement beyond great ape abilities, the ones which will surely play some prime-mover or hitchhiker role in the evolutionary scenarios that we try to construct for the last five million years. (Such "uniquely human" lists have proved very useful in the past - because they stimulate ape researchers to disprove them!)

This isn't, by the way, a list of "important" things. For example, dozens of fine primate studies over the last few decades have shown that we share a lot of social behaviors with chimps and bonobos (who have an amazing amount of hugging and kissing, reassuring touches, coalition behaviors, and so forth). These would make everyone's list of "important" attributes - and, forty years ago, such things would have made my up-from-the-apes list. They're obviously important to "being human" but - as it now turns out - they may *not* be order-of-magnitude advances beyond those of five million years ago. My list is a differential view (as it were, subtracting apes from humans), not the grand view.

Yes, that's a rather concentrated list, rather like a five-course French dinner stuffed into a small take-out box. My sympathies. But you get a week to dissect and digest it while I soak up some more linguistics and archaeology in a dim classroom in Paris, the variety with hard seats to keep the students awake. You can argue among yourselves about possible additions to, and subtractions from, my list. It's a game anyone can play.

SEVERAL MINUTES AGO, we were in farmland, and I was puzzled at why the other passengers were putting their coats on. Well, I just looked up again and I don't think we're in Kansas anymore. That's Paris out there. The Eurostar train isn't *that* fast. It's just that France has some well-delineated urban areas that lack the usual sprawl of many other cities, the strips extending from the core along the traffic arterials, so like the way that invasive cancers spread along arteries, veins, and nerves.

The French, at least in places, have learned to deal with that urban form of cancer. Maybe they will build berms someday to protect people in the cities from the train's noise, now that they have set a precedent with consideration for the cows.

Imagine a world without Darwin. Imagine a world in which Charles Darwin and Alfred Russel Wallace had not transformed our understanding of living things. What . . . would become baffling and puzzling . . . , in urgent need of explanation? The answer is: practically everything about living things. . . .

– HELENA CRONIN, *The Ant and the Peacock,* 1992

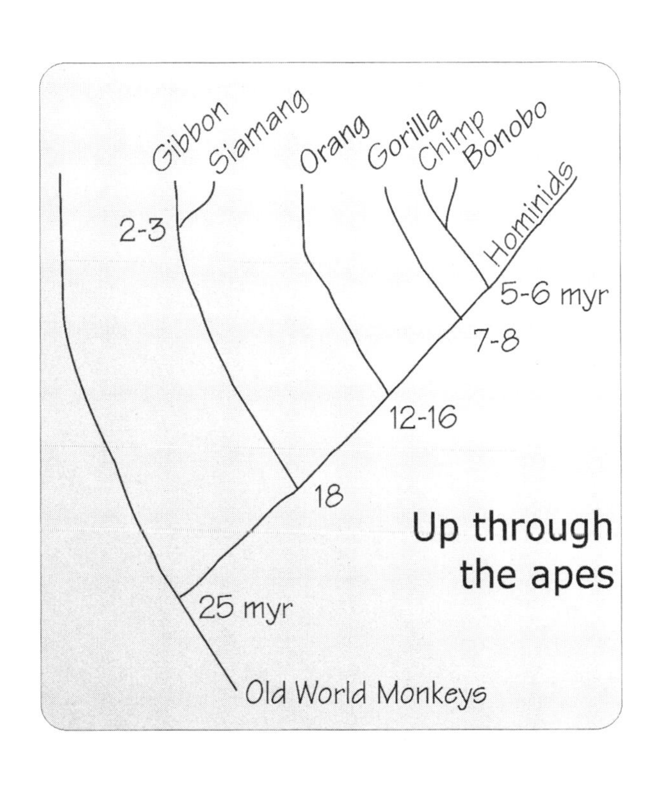
Gibbon
Siamang
Orang
Gorilla
Chimp
Bonobo
Hominids
2-3
5-6 myr
7-8
12-16
18
25 myr
Old World Monkeys
Up through the apes

To: Human Evolution E-Seminar
From: William H. Calvin
Location: 48.9°N 2.3°E
Musée de l'Homme in Paris

Subject: **The Ghost of Habitats Past**

After the meetings on the emergence of language broke up, an anthropologist friend and I headed over to Musée de l'Homme to see their excellent new *Homo erectus* exhibit, an inspired improvement on what used to be an endless display of stones and bones - the "hard evidence," if you like.

The general up-from-the-apes outline, for those who don't know it already, starts with apes evolving from the Old World Monkeys more than 25 million years ago. Gibbons split off about 18 million years ago. Great apes, at least since the common ancestor with the orangutans 12-16 million years ago, have had twice the brain of a similarly-sized monkey. The vegetarian gorilla, with the enormous gut and jaws to match, split off about 7-8 million years ago.

Chimpanzees are more omnivorous, with lots of ways of making a living. Our last common ancestor with the chimps and bonobos was 5-6 million years ago, and that tends to be the point of departure for hominid discussions. We start with something like a great ape, omnivorous but with a gut still accustomed to mostly fruit and leaves, not to major amounts of meat and fat. This apelike ancestor was capable of short bouts of walking upright, but not really adapted to sustained upright posture and efficient running. It likely had a lot of humanlike social behaviors, but not extensive cooperation and gossip.

There aren't many ape fossils from back then, despite a lot of "earlier than thou" competition among the paleontologists, but we tend to assume that the 5-6 million year creature looked and acted something like the chimp and bonobo. Certainly, if you want to see a living, breathing candidate for what our ancestors used to be like, spend a few hours at a zoo observing a dozen bonobos. The San Diego Zoo has an excellent exhibit and so does Planckendal (between Brussels and Antwerp). Bonobos have more behavioral overlap with humans in some areas (using nonreproductive sex as a social tool, for example) – though there are other areas (such as ganging up five-on-one to cooperatively murder a neighbor) where, alas, we seem more like the chimps.

Why the difference between such closely related species? It's probably nothing as simple as testerosterone levels but rather something more along the lines of social group size and structure. Even though young bonobo females tend to emigrate to another band in the familiar incest-avoidance strategy, they develop strong bonds there with other females and generally tend to dominate over males. Chimpanzee troops have much stronger "band of brothers" aspects. Despite some hostility between bonobo groups, peaceful mingling often occurs; in chimpanzees, hostility between groups is the rule.

Upright stance seems to have caused the first major anatomical modification from the common ancestor (though some behavioral modification likely preceded that). Whenever something major like that happens, there is often a whole series of new species soon thereafter, each somewhat different (it's called an adaptive radiation in evolutionary biology, and it's not unlike all the diverse web ventures of the late 1990s triggered by internet popularity, each dot.com trying to find a new niche to exploit). So one expects to find a number of different hominid species back at 5-6 million years ago, trying out the new niches that upright posture opened up. Most surely died out within a few million years or so.

In evolution, new behaviors routinely precede the appearance of concrete adaptations that facilitate those behaviors.
– ALAN WALKER &
PAT SHIPMAN, 1996

The initial stimulus for the upright rearrangement is unknown - maybe wading offshore, maybe carrying babies unable to cling to missing body hair, maybe foraging where the gathering had to be hauled home rather than consumed on the spot in the usual ape fashion. Certainly upright locomotion per se is initially very inefficient but, once the knees and hips rearrange themselves, it's probably an improvement on the knuckle-walking of the chimps.

At about the same time, there was a loss of forest as climate cooled. Unlike the other apes, our ancestors adapted to open woodland and then savanna, with all the attendant problems of competing for food with big savanna predators (and avoiding becoming their dinners). Likely the transition from savanna (grasslands with scattered trees) to the treeless steppes was far harder, what with not a single tree to serve as a perch or refuge.

Extensive Serengeti-style savannas didn't develop in Africa until about one million years ago, although fossils of browsing and grazing species did start to increase by 7 million years ago. Some lower-body characteristics of upright posture are seen by 5-6 million years ago, and all of the australopithecines seem to have lived near wooded habitats - or, at least, near the waterholes frequented by animals that did live in the woods. There's the occasional suggestion that there was a period when our ancestors might have waded among the flora and fauna near shorelines.

By three million years ago, when more savannas developed in the Rift Valley, the story must have involved following river valleys deep into Africa and adapting to the grasslands that border the waterways and lakes, with major behavioral adaptations for making a living in a way unlike forest-dwelling chimps. The widening and deepening Rift Valley likely helped to separate the chimp-bonobo species to the west from the bipedal apes of East Africa, in what Yves Coppens likes to call the "East Side Story." It perhaps gave them some room to develop separately, without competing constantly. Indeed, some monkeys also adapted to the East African savanna about the same time, and competition from such "baboons" would have been a factor – just as competition from the local monkeys, who can clear out fruit trees far faster than chimps, is a major factor for the surviving chimpanzees of Uganda. While the number of monkey species has been growing for millions of years, the number of ape species has been declining, proving that bigger brains aren't everything.

Our ancestors weren't leaving around much evidence of stone toolmaking, though they were likely as clever with found-object tools as their modern chimp cousins, all those termite-fishing wands and nut-cracking hammers. Their brains surely were functionally diverging from the chimp model, but their brain size was still in the great ape ballpark. The australopithecine teeth suggest they were eating a lot more rough stuff than chimpanzees, and the males were twice the size of females (in chimps and humans, males are only 10-20 percent larger). Both of these away-from-the-chimps trends reverse with the appearance of *Homo* about 2.4 million years ago. The alternative to them reversing is that australopithecines aren't really our ancestors – that some other species with small teeth and less oversized males is yet to be found in the ancient layers between 6 and 2.5 million years ago.

SOMETIME THIS SIDE of three million years ago, the climate drifted into the ice age jitters. Both chimp and australopith populations were surely downsized and forced into refugia (small isolated regions able to sustain the traditional way of life for a fortunate few). The ancestor of chimps and bonobos likely had refugia on each side of the Congo River. Within each refugia, they inbred and drifted and adapted to yield what we now see as the two surviving species, *Pan troglodytes* and *Pan paniscus*. Something similar likely happened in East Africa to the australopithecines, yielding a new variant which, starting about 2.4 million years ago, had a significantly larger brain. This was when we start talking of a *Homo* lineage with bigger brains (and with the cortical folds in somewhat different places, suggesting reorganization), unapelike inner ears, smaller teeth, and likely carrying around infants that were even more helpless at birth than are ape infants. The males likely weren't competing with one another for access to females quite as much as earlier.

I won't try to explain all of the species names that have been proposed for the first half-million years after the split. For one thing, they change unreasonably often - poor old Zinj has been renamed more often than the telephone dialing prefix for London has changed. But by 1.88 million years ago, *Homo erectus* is up and running - and running all over eastern and southern Africa. And, indeed, into Europe by 1.7 million years ago and all the way to southeast Asia by 1.6. It's the first "out of Africa" for the hominid lineage, though we still know little of its details.

By about 750,000 years ago, some new experiments were underway that led to Neandertals and modern humans, but *Homo erectus* persisted until about 50,000 years ago in China. Erectus was the longest-running species in hominid history.

Again, I won't try to disentangle the spread of transitional species that led from *erectus* to *sapiens* over a half-million years (the period from 0.75 to 0.25 million years ago may be another adaptive radiation, like the earlier one from 2.4 to 1.9 million years), but by somewhere between 175,000 and 125,000 years ago, there were anatomically modern *Homo sapiens* around Africa. They were in the Levant by 100,000 years ago (not surprising as the flora and fauna were often African). Somewhere before 50,000 years ago, they spread out into the rest of Asia (and soon thereafter into Europe and Australia) in the most recent Out of Africa expansion.

Did they interbreed with the indigenous populations they encountered (say, Neandertals or Asian *Homo erectus*)? They surely tried when populations downsized and there wasn't much choice in mates. It has been difficult to find regional genetic markers in modern populations suggestive of

"Neandertal genes," though some regional anatomical oddities during the last ice age certainly suggest mixed ancestry. This isn't an argument about different groups of modern humans. Everyone is related to everyone via common ancestors in Africa at that period; the issue is mostly one of how gradually the modernity transition occurred back in the last ice age.

Since this last Out of Africa, some major regional attributes have developed from the immigrating Africans. Roughly, they are now African, Asian (those that turned right

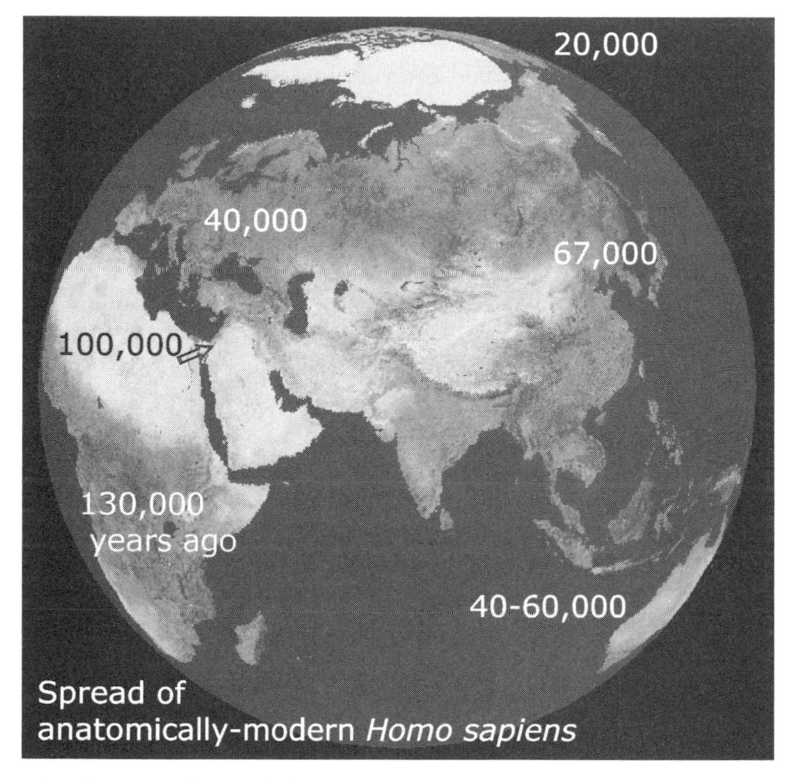

after leaving the Middle East), and European (those that later turned left) – though mixed types abound from later population movements, blurring differences.

And there is a lot of regional variation within each modern type, working against Platonistic "ideal" notions.

Variation in a population can make for strength in the long term, just as a mix of particle sizes is important in concrete and steel alloys. There is far more variation within modern African *Homo sapiens* than there is within the Out of Africa groups. Modern Africans really need the variation in reserve, just to work around the challenges from their parasite load and their fickle climate. The Asians and Europeans, besides being less rich in genetic variations that they can tap, seem to have specialized somewhat toward one end of the parental-care spectrum, concentrating on relatively fewer offspring (their biology results in having fewer fraternal twins) who grow up more slowly (somewhat slower growth rates, later puberty, and so forth).

AN IMPORTANT ASPECT OF EVOLUTIONARY HISTORY is the notion that there is a tree of species - and not, say, a web. Even real trees occasionally have places where branches lean against one another for so long that they stick and fuse (an "anastomosis").

Hybrids are not always sterile. Just imagine what it would do to the usual branching diagram of the higher apes if, for example, the chimpanzee lineage had been formed by a crossing of the gorilla and hominid branches. Our assumption of binary branching, in effect, says that branches once established do not recross.

This becomes a real problem when you know that hybrids are often fertile, as with crossings within modern human geographic groups. The history of "races" is full of such crossings and recrossings, a weblike history that makes one cautious of any treelike diagrams (Wisteria vines are more the model). Europeans are a mix of Asian and African from many times and places.

So comparing among the African-European-Asian "races" has its hazards. If you just classified using skin pigmentation, for example, you'd lump peoples together differently than if you paid attention to skull shape and body proportions, the way physical anthropologists tend to do. For example, the

"European" skull shapes are found in both the fair-skinned Scandinavians and the dark-skinned people of the Indian subcontinent. Evolution teaches us to expect a lot of small regional chance variations and adaptations, and we certainly see that humans are no exception to biodiversity. But we are all, not very far back, black Africans.

Already you can see that a lot more contributes to the hominid evolution story than merely stones and bones. We have a hard time evaluating the evolutionary history of breath control, for example, so important for things like diving and speech. Lots of soft materials like wooden spears didn't survive very well (though there are now some from Germany that are about 400,000 years old, balanced much like the modern javelin). We can now infer a lot about ancient climate and we're getting better at piecing together what ancient behaviors might have been like.

Maybe someday soon we'll even see the step up to higher intellectual functions (syntax, planning, logical chains, games with arbitrary rules, structured music, coherence-finding) more clearly. Intelligence was greatly augmented by the recent (maybe no longer ago than 50,000 years) evolution of higher intellectual function. Did the abrupt climate changes give our ancestors some opportunities that great apes didn't get?

WE'RE NOW TAKING an espresso-and-dessert break, though I suspect that wine or beer would be a better therapy for the sore feet that tend to come with viewing the hard evidence via even harder floors. We spent a lot of time discussing the so-called "bi-faces," ubiquitous stone tools that are flat and edged, from chipping away on both faces. Some have teardrop-shaped symmetry. Indeed, they are not only hand-sized, but they look like a hand held palm-out, with fingers close together.

A particularly fancy version is called the "handaxe," a term that has misled generations of researchers. They'd fit into the palm of your hand like a discus (some are small enough for a child's hand, others seem too large for anyone's hand) but most would be rather awkward as striking and cutting tools because of having sharp edges all around (you'd think that they'd have learned not to sharpen them all around, were pounding their main use). And no, they weren't hafted, to fit into a handle of some sort. As Richard Leakey once said of them, "embarrassingly, no-one can think of a good use." I'll say something more about them when I get to Kenya.

The restaurant here at the museum was designed to have a nice view of the Eiffel Tower, which is just across the Seine. The tower doesn't distract me a bit from thoughts of *erectus* ancestors. But we're looking down on a river. And around the tower is a flat expanse, studded with grass and trees. Compared to the surrounding city, it's almost savanna.

IF I IGNORE THE REGULARITY of the walks and plantings across the way, it reminds me of what my colleague Gordon Orians said about views that make people feel good. As a behavioral ecologist, he speaks figuratively of the Ghost of Dangers Past (we dream of spiders and snakes, not current dangers such as cars and handguns). He says our human aesthetic sensibilities are similarly influenced by the Ghost of Habitats Past. Habitat selection by an animal is influenced by where it grew up, by where it sees others of its own species, and - especially when those criteria aren't working very well - by some innate knowledge of what the species' former habitats looked like. There's no reason that humans should be an exception.

As E. O. Wilson likes to summarize Gordon's results in his sociobiology lectures, a high-ranked vista for humans generally includes some water (stream, pond, seashore). A forest view isn't as good as one with some scattered trees (not too tall, either; trees that spread out in horizontal layers like acacias get higher viewer ratings). A few large animals in the distance (but not too close for comfort) is an attractive option. And, for best effect, the scene should be viewed from a slight elevation, preferably framed in a way that suggests viewing from some shelter. The view from the restaurant qualifies.

In short, I would conclude, it's the view from a tree nest in our ancestral savanna home. Such gut feelings tell us something about our ancestors - indeed about what they liked to put in their guts. Such innate likings would have guided individuals in selecting a habitat suited to the better ways of making a living for their species, back then - telling them when to settle down, when to move on to "a better view."

Oriental landscape architecture adheres to this savanna-tree-house formula, what with that little shelter on the artificial hill from which to survey the ponds and scattered trees. It's species specific to us humans - a chimp or bonobo would have a different esthetic, likely featuring more of an inside-the-forest view of fruit trees. They might find our open spaces threatening.

But redesigning the Eiffel Tower area, to be even more like New York City's Central Park (a nicely-designed open woodland with pastures and ponds), gets ahead of the story. I'll be in the Rift Valley soon, so let me save tree-house esthetics until then. Maybe this belongs on the hominid bootstrap list, if we can ever figure out chimp esthetics as a basis for comparison.

SO WHY DID SO MANY THINGS change in the same five million years? Some, of course, probably changed as a group, thanks to sharing some common neural machinery in the brain. When you improve one bit of anatomy, you sometimes make possible another seemingly unrelated function.

It's much like when wheelchair considerations paid for curb cuts but soon 99 percent of their use was for things that would never have paid their way - baby carriages, skateboards, wheeled suitcases, bicycles, and so on. Maybe one of those secondary uses will eventually pay for further improvements, but the "free lunch" is alive and well in both urban architecture and biology.

For a local example of how shared-use structures can serve as a stepping stone to another specialized form, just climb the circular stairs inside the north tower of Notre-Dame and look around. Back in the days before downspout plumbing, rainwater from the roofs stained the sides of buildings, so they invented horizontal rain spouts to carry rainwater a meter or so out from the side of the building. Then they began to decorate the rainspouts with animal-like features - so they became multi-functional, good for both plumbing and as a prophylactic for warding off evil spirits. From these shared-use versions, downward-looking gargoyles evolved, ones without rainspout functions remaining. Then upright gargoyles evolved, perched on ledges and no longer looking like potential rainspouts at all. The chimeras at Notre-Dame are particularly famous examples, from about 800 years ago. And maybe our unusually-capable brains are too, from a somewhat earlier era.

The higher intellectual functions (syntax, multi-stage planning, structured music, chains of logic, games with arbitrary rules, and likely our fondness for discovering hidden patterns) may all share some neural machinery, as I have often discussed. Maybe when you improve structured language, you get better at structured music "for free," without having to have separate natural selection involving four-part harmony.

Derek Bickerton argued, in our *Lingua ex Machina* book, that altruism needed some abstract mental categories for things like giver and recipient. That could have helped set the stage for structured language. And I argued there that the "get set" planning for ballistic movements like hammering and throwing needed mental machinery that would have found a secondary use in sentence structure, all those embedded phrases and clauses. Once you can categorize for reciprocal altruism and plan for ballistic movement, maybe you can do the structured planning on different time scales, using the same parts of the brain to plan an agenda or a career.

I think that this focus on shared neural machinery is far more useful than the usual bigger-must-be-better focus on brain size. Everyone just assumes that bigger brains were a good thing and, while I think that's likely true in some sense, I find myself playing the skeptic.

Again, this is a game where anyone can play (though relevant knowledge always trumps speculation!), so have fun discussing it while I'm traveling and out of touch. Another stop in Europe, and then on to Africa.

To:	Human Evolution E-Seminar
From:	William H. Calvin
Location:	49.6°N 8.2°E Bockenheim, Germany

Subject: **Tracing our roots back to the Big Bang**

I'm at a family reunion near Worms, in the German wine country. And there is a winery in the family. I have to give an after-dinner speech, and so must abstain from sipping the excellent wine. I hope the Liebrich audience likes the following, because I mostly had this e-seminar audience in mind when writing it. It's a quick sketch of what happened when.

[Clearing throat.] The Liebrich that emigrated from Mannheim to Pennsylvania in the eighteenth century is about six generations back for me, on my mother's side of the family. Just as you have two parents, four grandparents, eight great-grandparents, 16 g-g-grandparents, 32 g-g-g-grandparents, so you have 64 g-g-g-g-grandparents. Thus I have 63 more family reunions like this one, yet to attend. For the younger people here, it is eight generations back. That ancestor eight generations back represents only a fraction of one percent of all your genes. You have 255 more reunions to attend, to give equal time to all your ancestors of 200 years ago.

This is unlikely to happen, of course, because it is so rare for someone to have the energy and enterprise to organize an event like this [here I thank Dr. Winfred Liebrich of Berlin for organizing the affair].

So, what was the world like, several hundred years ago in the year 1800? The Napoleonic Wars were involving all of Europe, and the French had annexed this vineyard-rich left bank of the Rhine.

We were almost totally ignorant of how the brain worked in 1800, with little advance over what the ancient Greeks and Egyptians guessed. Medicine in general was primitive in 1800 (bleeding and purging were in fashion). Not only hadn't surgical anesthesia been invented yet, but doctors hadn't learned to wash their hands and were spreading disease from one patient to the next. However, vaccination for smallpox had just been invented by the English physician Edward Jenner in 1796.

In the physical sciences in 1800, there was a remarkable scientist here in Germany called Count Rumford. His is a case of emigration in the other direction. Benjamin Thompson was born into a Massachusetts farming family, and worked his way up to European nobility. His politics placed him on the losing side of the American Revolution, and so in 1776 he moved to England and eventually became Sir Benjamin Thompson. It was later, when the Duke of Bavaria asked him to come to Munich to reorganize the army, that he became Count Rumford of the Holy Roman Empire. He invented things from stoves to welfare systems. A very versatile fellow, somewhat like Benjamin Franklin but considerably more abrasive.

As a scientist, he predicted in 1797 that the cold water in the ocean depths had to be coming from Arctic latitudes, a cold deep counterpart to the warm surface Gulf Stream. It's a river under the ocean, flowing *southward* down near the *bottom* of the Atlantic; it plays a big role in our modern understanding of abrupt climate changes, as it causes the warm northbound Gulf Stream to fail occasionally.

Twice as far back, in the year 1600, is sixteen generations away (for humans, a generation averages about a quarter-century). If you claim Shakespeare as an ancestor, remember that your relationship is only 1 part in 65,000 and that any such genes might well have been lost by chance, so that you are only a relative on paper. Such is the power of dividing by half every quarter-century.

Dutch opticians invented the telescope in 1600 and soon Galileo was busy reinterpreting the heavens – a risky business. Giordano Bruno, the wandering monk who influenced Spinoza and Liebniz, was burned in 1600 as a heretic, for (among other things) speculating about life on other planets, much like scientists do today.

Our genes may be immortal but the collection of genes which is any one of us is bound to crumble away. Elizabeth II is a direct descendant of William the Conqueror. Yet it is quite probable that she bears not a single one of the old king's genes. We should not seek immortality in reproduction.

But if you contribute to the world's culture, if you have a good idea, compose a tune, invent a sparking plug, write a poem, it may live on, intact, long after your genes have dissolved in the common pool. Socrates may or may not have a gene or two alive in the world today . . . but who cares? The meme-complexes of Socrates, Leonardo, Copernicus, and Marconi are still going strong.

–RICHARD DAWKINS,
The Selfish Gene, 1976

Now let's jump twice as far, 800 years back to the year 1200 when building cathedrals was in fashion. By this time, hay had been invented, and storing grass for the winter made it possible for farmers to maintain large herds of cattle instead of slaughtering most of the herd every autumn. That improved meat supply is credited with allowing substantial cities to finally develop in northern Europe, whereas they had formerly been restricted to sites nearer the Mediterranean. It's also the medieval warm period in Europe, but that's about to end. Already climate change is making life miserable in the American southwest with centuries-long droughts, and by 1300 Europe slipped into 550 years of the Little Ice Age.

Skipping back 1,600 years takes us to the year 400, when the Roman Empire was falling apart. It's the time when the Huns destroyed Worms, just east of here.

At 3,200 years ago, in the year 1200 B.C., we're talking Old Testament times, complete with pharaohs. Island cities, such as those on Crete, weren't fortified then because sea-borne warfare hadn't yet been invented. We've skipped over those great centuries in Greece between 500 and 200 B.C. when

Pythagoras inferred basic geometric regularities and musical chords, Empedocles talked of the survival of the fittest in evolution, when Aristotle, Plato, and Archimedes created the classical foundations of philosophy and science.

At 6,400 years ago, we are back to the first cities. Farming settlements had been around for 5,000 years to the east of the Mediterranean, but not real cities with a lot of specialized occupations far removed from agriculture, like tax collectors. Written history goes back only 5,000 years, which is when writing was invented in Sumer to keep tax records.

Going back 12,800 years ago lands us at the origins of agriculture and animal domestication. It is when the Mediterranean islands were first inhabited, suggesting sailboats at the least. Things had been warming up out of the last ice age, starting about 15,000 years ago. It had melted all the ice sheets in Scotland, and the Scandinavian ice sheet was down by half. But 12,900 years ago was the time of a big surprise, around most of the world. Suddenly, just in the matter of a decade or so, the climate flipped from warm-and-wet into the cool-and-dry mode, with temperatures plunging back to what they had been in the ice ages.

Here in Germany, the forests disappeared and vegetation characteristic of modern Siberia took its place. This lasted for over a millennium until, even more suddenly, it warmed back up again and the rains returned. This down-and-up event, called the Younger Dryas, is only the most recent of dozens of similar flips between warm-and-wet to cool-and-dry, usually recurring every several thousand years.

At 25,600 years ago, we're back in the coldest part of the last ice age. Needles made of bone had just been invented. North of here was mostly ice: a giant mountain of it sat atop Scandinavia – that's why northern Germany and Poland look so flattened – and another ice sheet sat atop Canada, as tall as a mountain range. Sea level then was at its lowest, forty stories below where it is now. But it is also the time of the cave paintings such as Lascaux, a period when our ancestors clearly had acquired a modern suite of mental abilities - they thought, and communicated, much as we do. Earlier than about 50,000 years ago, we're not so sure of that, even though people then had modern bodies and brain size.

By 50,000 years ago, we're back before cave art. Behaviorally modern *Homo sapiens* (people like us) were in East Africa (this is when you start to see the first evidence of fishing) and by 40,000 years ago, they were spreading westward into Europe from somewhere in Asia, encountering the Neandertal peoples already living there on the ice age frontiers. Shortly afterward, the Neandertals were mostly gone and people like us were the only hominid species left on earth, even spreading into cold, arid places where the Neandertals never lived. *Homo erectus* was still in China but modern humans had recently arrived in Australia, showing that they had mastered water travel, to make it across from southeast Asia.

Going back to 100,000 years ago, at least three hominid species were around. People like us were, however, only in Africa or nearby (and all of the Liebrich ancestors had black skins, too). Modern *Homo sapiens* had probably been in Africa during the last warm period in the ice ages, which started 130,000 years ago (this is when you see the first use of fireplaces as a centrally-located feature of encampments, suggesting some change in social organization). The warm period lasted until 117,000 years ago, when things abruptly cooled, much as in the Younger Dryas - but it stayed down in the cool-and-dry mode. Major ice sheets didn't develop until about 70,000 years ago, perhaps helped along by a major volcanic eruption in Indonesia that reflected a great deal of sunlight back out into space.

Doubling again, back to 200,000 years ago, and we're into the prior ice age (they last about 100,000 years between major meltoffs, and there have been dozens of them). People like us were probably *not* around then, just various large-brained but ruggedly built people with brow ridges.

Back at 400,000 years ago, there were a variety of confusing hominid species called Neandertals and "archaic Homo sapiens," not only in Africa but in Europe and Asia. *Homo heidelbergensis* might be the variety which was our ancestor.

At 800,000 years ago, the ice age rhythms change somewhat, the major meltoffs shifting for some reason from a 40,000 year interval to our present 100,000 year interval. Among our probable ancestors, there is *Homo antecessor* in the Iberian peninsula and *Homo erectus* in Asia and Africa - but a variety of variants are starting to appear, some of which later led to archaic Homo sapiens and Neandertals.

At 1.6 million years ago, *Homo erectus* is in both Africa and southeast Asia, and the smaller-brained australopithecines are also running around Africa (they die out by 1.0 million years ago, leaving *Homo erectus* the only hominid around).

Back at 3.2 million years ago, there were only australopithecines ("southern apes"), and only in Africa. While upright in posture, they only had a brain the size of an ape's, about a third the size of our brains. While upright, they were still living pretty close to trees. But momentous things are happening, as the earth is about to enter the ice ages after millions of years of a cooling and drying trend that has started to create savannas in Africa's Rift Valley.

North America and South America were not connected back then, and tropical ocean currents could still flow between the Atlantic and the Pacific Oceans in what I like to call the Old Panama Canal. But Panama rose out of the seas, slowly damming up the passage. By about 3.2 million years ago, the ocean currents were forced to rearrange themselves in a big way. That seems to be the best candidate for the final event that tripped the ice ages, which began in earnest between 3 and 2 million years ago. That's when the *Homo* lineage split off from australopithecines, brains got bigger, and toolmaking started in earnest.

Now we're back to 6.4 million years ago, and that's about when we shared a common ancestor with the modern chimpanzees and bonobos. The hominid lineage split off then, and was already upright in posture at about 6 million years ago. Bipedal apes, no less. We're really the third chimpanzee, as Jared Diamond likes to say.

Going back to 12.8 million years ago, we're near the common ancestor with orangutans (gorillas split off in the meantime, about 8 million years ago).

Back to 25 million years ago, apes evolved from the Old World Monkeys.

And at 50 million years ago, we're all still monkeys getting larger after the big extinction at 65 million years ago which killed off the dinosaurs.

At 100 million years ago, we're some lower form of primate, no larger than a squirrel but smarter. Dinosaurs rule the earth, not our ancestors.

Around 200 million years ago, we're early mammals, even smaller and less significant, trying to survive the Permian extinction.

Back at 400 million years ago, our ancestors are just venturing out of the sea onto land as some sort of lungfish or amphibian.

At 800 million years ago, we're in the Precambrian, and all the evolutionary action is still in the seas and very small. It's multicellular by now and looking quite weird when hard enough to fossilize. A truly shocking event happened to the earth's surface several times about then: it froze solid, from north pole to south pole, white all over. Volcanos peeked through, however, and their carbon dioxide provided the greenhouse gases that eventually rewarmed the earth and melted back the ice. Life in the sea wasn't entirely frozen, as there were surely hundreds of pockets near the volcanoes which remained above freezing. This isolation was one of the setups for the great Cambrian explosion of life that occurred 530 million years ago.

Around 1.6 billion years ago, sex was being invented as a way of improving on bacteria-swapping genes occasionally, the primitive gene-mixing mechanism. A new committee of cell parts had evolved, called the eukaryote, with the genes kept in a bag in the middle of the cell, called the nucleus. This advance beyond the bacteria was momentous, allowing cells to be much more complex and capable.

But back at 3.2 billion years ago, there were only bacteria-like cells around, no committee-like cells at all. Life evolved 3.8 billion years ago in the oceans and around hot vents.

Go back 6.4 billion years ago, and the Earth didn't even exist (it coalesced at about 4.6 billion). At 6.4, our ancestors were just dust swirling around in space, fragments of a star that had exploded. Our "local supernova" about 7 billion years ago produced nearly all of the atoms presently in our bodies, cooking the heavier ones out of the hydrogen and helium that had been around for a long time.

Double the time again and we're back to the Big Bang at 13 billion years ago, not even atoms and molecules but just the very hot "quark soup" of the more primordial building blocks.

And you can't go back any further than that. So far as we can tell, that's when Time Began. That's when all the parts were put in place that eventually became Liebrichs (and bacteria and plants, too). So that's the beginning of the family tree, and the end of the talk.

AFTER THE FAMILY REUNION broke up the next day, each car departing with a case of wine in the back, we sat in the shade talking about my departure the following day for Africa. My cousin with the vineyard got to asking me questions about brain size. Did the *Homo erectus* brain really increase in a stepwise fashion, he asked via our weary interpreter, or was it gradual? Maybe a series of small steps, I answered, hedging.

And was it really true that brain size increased to make us more intelligent? It would be so easy to nod agreement. It's the conventional wisdom, the current default answer. It's probably true. And I cannot confirm the opposite, or say that brain size was really about something else. He's not in the field and sensitive to nuance, nor the student of the subject that this e-seminar tends to attract. But he's an intelligent man and deserves a better answer than "Ja."

There are great costs, I said, to increasing brain size. Our present brain is only 2 percent of the body by weight, but it accounts for 16 percent of the basal metabolism (the brain share is 3 percent in an average mammal, and some marsupial brains get by on less than 1 percent). Were our brain only a third its present size, it would take a lot less blood.

All of that additional blood flow had to come from somewhere in evolution, and some think that it was only after some high-calorie meat was added to the diet that gut length could decrease and free up some capacity. Another expense is that bigger brains take longer to grow up. They also decrease the efficiency of upright walking because of the birth canal size spreading the hips and promoting a return to the chimpanzee-like waddle.

Yet I'm not sure what's on the positive side of the balance. Yes, brain size increase could have been for intelligence in general. But something doesn't fit yet. So I'm not at all sure what bigger brains have to do with hominid evolution. I find myself asking questions like:

- Was the size increase really required? (In the sense that a chimp-sized brain couldn't possibly manage all those augmented functions like syntax, not any more than you could run Windows programs on a 1950 desk calculator.)
- Or was it simply permissive? (In the expanding-economy sense, that it's easier to experiment with chancy new functionality if you don't, at the same time, have to compact or eliminate some existing function. These days, that's what is blamed for "RAM bloat," the escalating RAM memory requirements for new versions of software.)

- Or was brain size, per se, not involved at all? (Except as a minus. Size could just be a byproduct - reorganization of the brain might be the main thing, and the genes for reorganization might have had the side effect of increasing size at the same time, not being versatile enough to hold size constant while rearranging things). Or size and reorganization might have been late secondary consequences of something like bipedalism's rearrangements of body and brain.

My guess, for what it's worth, is that brain reorganization could proceed more easily in the bigger-brained variants of any generation, just because of more room to maneuver - but that size per se wasn't the name of the game. There were certainly no new modules tacked on to human neuroanatomy that the great apes lack, not so far as we can tell. However the brain is doing language, the areas involved seem to have "kept their day job," as the psychologist Elizabeth Bates likes to quip, while doing language as a second job. The same thing likely applies to those areas at the bottom of the frontal lobe which seem to have acquired the ability to maintain mental checklists and monitor progress on an agenda.

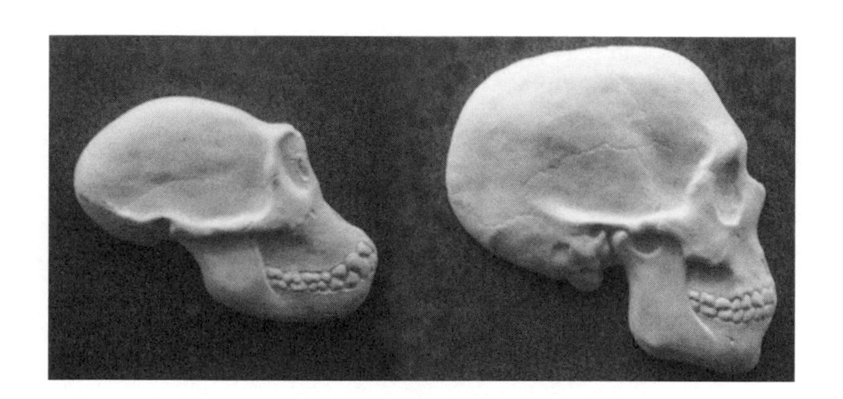

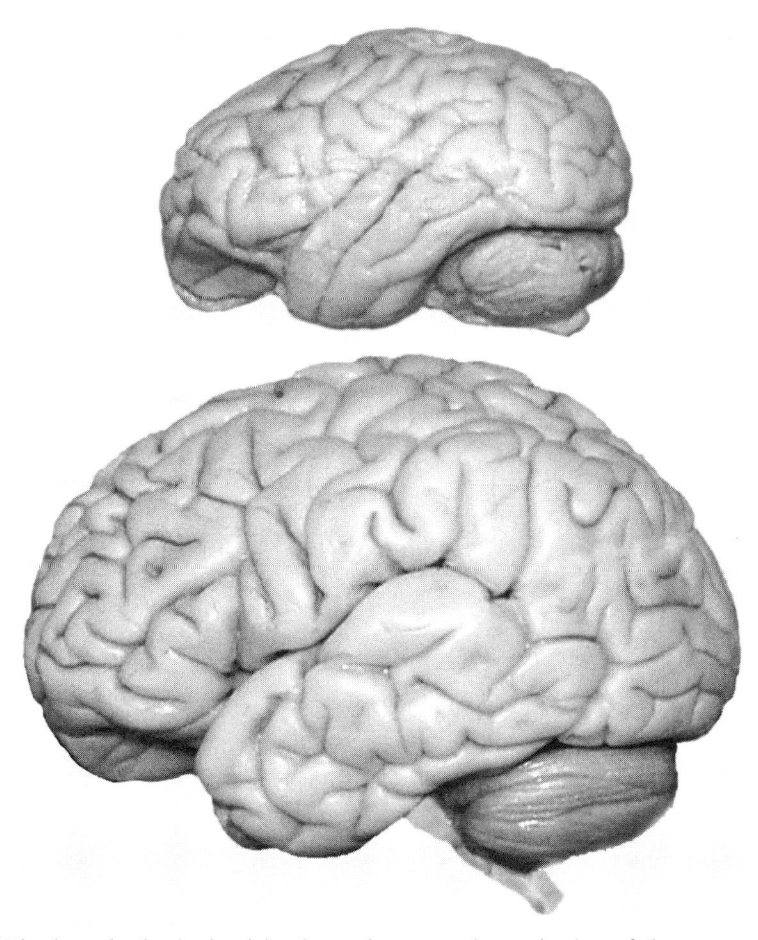

The bonobo brain (top) is about the same size as brains of the bipedal apes, such as the australopithecines.

The modern human brain (below, shown to the same scale) is about three times larger, with more than four times the amount of neocortical surface area. (Photographs thanks to Terrence W. Deacon)

To: Human Evolution E-Seminar
From: William H. Calvin
Location: 52.309°N 4.767°E -2m ASL
Layover Limbo

Subject: **IQ and evolution's package deals**

One common assumption about brain size is that if evolution acts on discrete abilities, perhaps the brain enlarges bump by bump - that the visual cortex bulges out when visual acuity comes under selection, then the motor cortex if dexterity comes under selection, and so forth. Something like this may be true for smell - looking at mammals broadly, you can enlarge olfactory areas without necessarily enlarging much else - but "enlarge one area, enlarge them all" seems a bit closer to the truth for neocortical areas. So if visual acuity comes under selection, so that more finer-grained neurons are needed, requiring more space for visual cortex, the whole neocortex may enlarge and so you may get more auditory cortex as a side benefit.

Much of the interest in brain size comes from a commonplace assumption, that bigger is better. If the fourfold increase from the apes to our average brain size (about three pints, 1350 cc) was so useful, maybe (so the assumption goes) bigger-brained people are also smarter. I have certainly come to be skeptical of this latter assumption, about how people vary.

Variation between species and variation within a species are two very different things, as it turns out. It's not that I doubt the data - IQ and brain volume aren't that hard to measure and there is some correlation between IQ and brain size in modern humans - but I'm cautious about the cause-and-effect presumption usually involved in interpreting the data. Bigger feet may correlate with height, but we don't usually assume that having bigger feet makes us taller. Instead we assume they're just both consequences of some other aspects having to do with nutrition, number of childhood diseases suffered, and heredity.

I also know that fat (in the form of the myelin which insulates the long wire-like parts of nerve cells) is a big part of brain volume, and that no one has yet evaluated whether the bigger-within-our-species brains are just fatter, much like two computers which differ only by the larger one having thicker insulation on the wires. We should soon be able to measure how much of brain size variation is just slender versus padded styles in brain fat and how much is processing power (and how much of that is in neocortex, the only part relevant to most innovative aspects of intelligence).

I'm also skeptical because intelligence (in the general sense of the word) simply isn't what is so easily measured with pencil and stopwatch. IQ does correlate with the "quickness" which is part of our general impression of someone's intelligence. Speed of decision making and how many abstract concepts you can juggle simultaneously (major aspects of what IQ measures) are undoubtedly important for being a modern physician but that's surely a peculiarity of the profession (and, more generally, of our modern survival-of-the-fastest society), not a general trait of goodness and human environments more generally. So don't conflate IQ with intelligence. Much of the practical side of intelligence has to do with being innovative in dealing with social and environmental challenges, and on longer time scales than a 10-minute office visit.

Note that the groups so renowned for practical intelligence are not the somewhat bigger-brained groups. (Asians lead, followed by Europeans and other mixed-race groups, but we're talking of differences of about two percent in average cranial capacity between Asians and Africans. This is splitting hairs.) Consider the environmental challenges of the Australian outback and the fickle climate in which Africans have thrived.

You see a lot of assumptions about bigger brains having been important for colonizing Eurasia with its wintertime challenges, but that's simplistic. I certainly suspect that the different environments of Eurasia caused some variants among the African immigrants to thrive better than others (and planning ability is often needed to get through the winter), and I'm quite willing to assume that the somewhat bigger brains came along with a package, but I'd really like to know what that package is.

ONE OF THE FEW CANDIDATES thus far is the so-called r-K spectrum of parental investment, well studied in the animal world and surely applicable to humans ("r" stands for the lay-them-and-leave-them parental strategy, "K" for the opposite extreme, heavy investment in relatively few offspring). "K" usually involves slower growth, delayed sexual maturity, and longer life span. Being able to afford only one or two children because of the thirty-year expense of putting them through higher education is often offered as the new extreme example in "K" but r-K is mostly a biological thing, seen in things not normally under voluntary control, such as having twins or not - and some animals switch strategies when the climate improves to cut corners (taking chances by having twins, weaning sooner, and so forth). Humans are at the far end of the animal world's r-K spectrum but we vary, with some not quite so extreme as others, particularly when boom times start suggesting corner-cutting possibilities.

But how is r-K implemented mechanistically? It's probably a package deal, just like you can't get power windows on your car unless you also get leather seats. They just aren't customized piecemeal, as much as customers would prefer it. With such a package deal in biological bodies, only some aspects need to have immediate payoffs in dealing with the environment or social life. Just because I opted for power windows, you can't infer that I like my leather seats (I'd much prefer cloth, cooler in the summer and warmer in the winter). They were merely dragged along.

So too, some things are dragged along in biological bodies – maybe even bigger brains, all without brain size having been one of the important aspects under natural selection. Genes just don't do things piecemeal, despite our tendency to name them as if they did; they too involve package deals. Many aspects of r-K (including brain size) might simply flow from slowing down the overall pace of prenatal and childhood development, or from decreases in overall testerosterone levels – global things, not piecemeal customization.

It's easy to let the bigger-is-better assumption keep you from thinking more deeply about this foggy subject, but as the view clears we may discover that brain size was just swept along in a general flow of other, more immediate, things. And that intelligence is mostly an aspect of another kind of package, those curb-cuts of secondary use, where the whole suite of higher intellectual functions profits from improvements paid for by more restricted uses.

WHAT WOULD CONSTITUTE a satisfying explanation of brain size? We'd like to know what happened when brain size took its larger steps up. We'd like to know how the course of development changed, likely via disturbing the regulation that held brain growth to the ape standard. (That's probably done by deleting or inactivating a regulatory gene, not necessarily by adding a new "big brain gene.") We'd like to know what variety of natural selection paid the price of admission. (Likely a series of them, different package deals at different times.) We'd like to know what curb cuts were involved, what new secondary uses were facilitated. And where.

You can easily double that brain-size list. Try another list for intelligence, while you are at it.

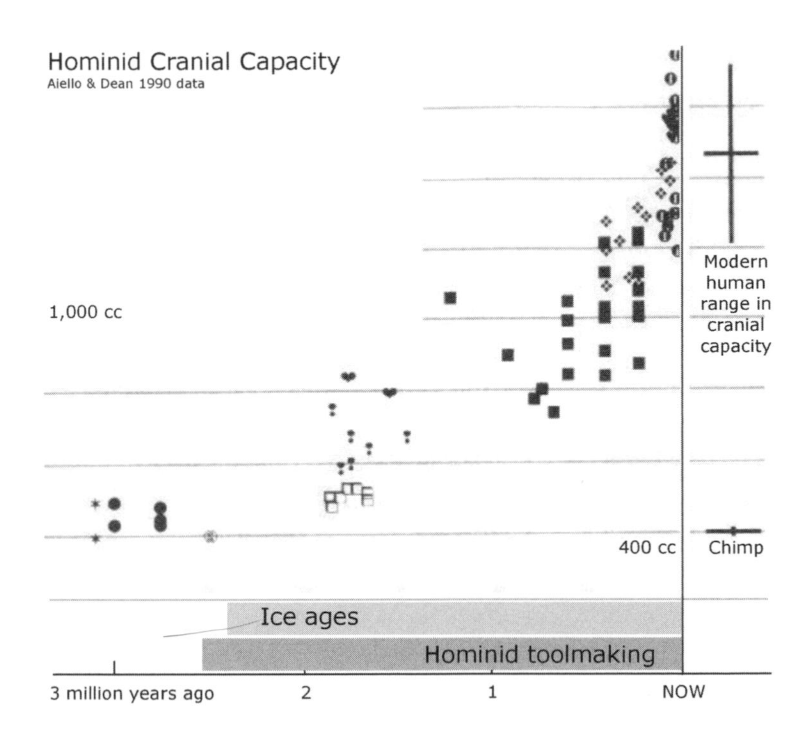

Hominid Cranial Capacity
Aiello & Dean 1990 data
1,000 cc
400 cc
Modern human range in cranial capacity
Chimp
Ice ages
Hominid toolmaking
3 million years ago
2
1
NOW

The Sahara has two potentially stable states: desert and heavily vegetated. It could flip between the two, and only a little more increase in rainfall could do it.

– VICTOR BROVKIN, 2001.

To: Human Evolution E-Seminar
From: William H. Calvin
Location: 22°N 14°E 10,000m ASL
The Sahara

Subject: **Why climate can suddenly flip**

The Sahara down below gets almost no rain at all. It doesn't even get dew from offshore fog drifting inland, like some deserts near a coast. It is "hyper arid."

Not always, however. There have been "pluvial" periods when the Sahara got enough rain. Between about 14,800 and 5,500 years ago (except for the Younger Dryas), it was a verdant landscape covered with grasses and shrubs, with numerous lakes. There were grazing animals of many kinds, even elephants.

There have also been periods in the ice ages when the arid area was even larger than at present. So why is there a Sahara at all?

Well, we're into the "horse latitudes," those bands of fickle winds and dryness that surround the globe near 30° North and 30° South. Lack of vegetation makes them brighter-looking. The Sahara is an example (the arid band actually extends east across Asia), and the Southern Hemisphere has the Kalahari and Australian deserts, plus Patagonia.

If hot air tends to rise, then what goes up at the equator has to come down somewhere else. All of those tropical rains are because the moisture drops out, once the dew point is reached during the ascent to cooler levels. By the time the tropical air comes back down from the stratosphere hereabouts, it is dry. These are examples of what the atmospheric scientists call the Hadley Cell circulation, named after the 1735 analysis by the British lawyer George Hadley (scientists used to make their livings in more diverse ways).

It is now known that each hemisphere is divided into three cells: rising air at the equator falls between 20° and 35° North (creating the Hadley Cell). Air rises again at roughly 55° to 60° North and descends over the North Pole (creating the Polar Cell). In between the descent at about 30° and the rise at about 60° is the third one (called the Ferrell Cell after a nineteenth-century American meteorologist). All of this varies with the season. Ditto for the Southern Hemisphere. This six-cell general circulation pattern is one of the reasons why the North and South Poles are so dry, as they are being flushed by moisture-free air that descends from on high, just like the Sahara.

They are, of course, high-pressure areas. In low-pressure areas, air rises and any moisture may precipitate out when the dew-point temperature is reached. Thunderheads are vast upwellings and can carry some heavier molecules (like refrigerator coolants) into the upper atmosphere; they'd never diffuse there on their own, but they go with the flow, another one of those package deals like brain size.

Another consequence are the bands of fickle winds ("the doldrums"), cursed by sailors for centuries, that occur near the equator and at the horse latitudes. They are because winds tend not to cross between cells, thanks to the vertical curtain of air separating adjacent cells. At the horse latitudes, sailors also cursed the relentlessly sunny skies (few clouds) and dry air, along with the lack of reliable wind to carry them out of the situation. Even the ocean surface is more salty in the horse latitudes, because it doesn't get rained on. The next time you walk through one of those building entrances with an air curtain rather than a door separating indoors from outdoors, remember the cell boundaries of the earth.

WHILE HORIZONTAL WINDS MAY NOT OFTEN PASS through the vertical curtains, the curtains themselves create the important "trade winds" and "westerlies" on which sailors and weather forecasters rely. To understand this, recall that even if you are standing still at the equator, you are moving eastward at a speed of about a thousand miles per hour (it takes 24 hours to rotate through the 24,902-mile equatorial circumference). But halfway to the North Pole at 45°N, your daily path is about 70 percent of the equatorial circumference, and so your eastward speed is 30 percent less (at 60°N, it's down by half). Same thing applies to the air, which is dragged along at the same local speeds by the surface features.

Now consider what happens to a wind blowing north. It also has a certain eastbound speed which it doesn't lose (conservation of angular momentum and all that) as it goes north, a speed greater than the local eastbound ground speed. So this northbound wind will also move east relative to the ground. It will seem to turn right. Try to move north and you really move northeastward.

Now consider the fate of the air that descends from on high. Air that descends in that vertical curtain at 30°N and turns north continues to travel east at the velocity characteristic of 30° - even though its surroundings are now traveling eastward more slowly. This northeast-bound air stream will thus appear as a wind coming out of the southwest to a local observer. These southwest winds get called "westerlies."

Air from the 30° descending curtain that turned south will be traveling eastward at the 30° velocity but, in this case, their surroundings will be traveling faster. So these winds will appear as a wind out of the northeast to a local observer, as if they too had turned right. They got called "trade winds" not for commerce but because they were so steady, "threading" along at a constant pace.

This is, you may have guessed by now, the so-called Coriolis effect at work (it's not a "force" so much as just conservation of momentum). It appears to turn moving things to the right in the Northern Hemisphere, and to the left down south. That's what George Hadley figured out. Sailing ships heading to North America from Europe went south past the doldrums to pick up the trades, but returned on a more northern path using the westerlies.

JUST SCANNING THE GLOBE FOR DESERTS, you can see exceptions to the 30° ideal all around 30° North and 30° South. (Florida would be a desert were it not surrounded on three sides by the Gulf Stream.) And the Ferrell-to-Polar Cell boundary at 55-60° North is something of a statistical thing, not exactly a vertical curtain, and there are eddies (alias weather systems) that wander around.

It's also not clear that things have always been the six-cell way, or that this pattern must continue. Maybe there are two-cell possibilities or no-cell chaotic arrangements. Any such reorganization of this cellular circulation would, of course, have profound all-bets-are-off consequences for regional climates and the world's average temperature. No more steady trade winds or westerlies.

This is not like the more familiar droughts, where your rain happens to fall elsewhere for a decade (and so others prosper for awhile). These reorganizations last a long time and have global consequences. The best-studied case so far is the abrupt warming in the Northern Hemisphere about 15,000 years ago that started the ice sheets to melting. Prior to that time, Lake Victoria had dried up for lack of sufficient rainfall; abruptly, it got enough rain once again. Prior to then, there were big lakes in Nevada and western Utah; abruptly, they dried up and became as arid as today. Such simultaneous occurrences in distant places are why people think that the atmospheric circulation pops into a new mode of operation.

Even if you conveniently happened to live in a place where the local climate changes balanced out, reorganization would likely affect global *average* climate as well and so you'd still suffer from the worldwide flip from warm-and-wet into cool-and-dry. That's because reorganization likely wouldn't maintain the same average amount of water vapor in the atmosphere (it's the most important "greenhouse gas," not carbon dioxide or methane). The whole earth could get warmer and wetter, or cooler and drier, in just the few years that it would take the winds to rearrange themselves. Oceans are usually thought to take much longer to change (it takes six years for ocean currents to get from Labrador to Bermuda), so those who worry about abrupt climate change tend to look at potential cell reorganizations in the atmosphere with considerable interest. The other quick effect is albedo, with a brighter atmosphere (from dust storm or salt spray) bouncing more sunlight back into space.

Whereas the atmosphere is quick and agile and responds nimbly to hints from the ocean, the ocean is ponderous and cumbersome.
– GEORGE S. PHILANDER,
Is the Temperature Rising?, 1998

Apropos causation, remember the difference between proximate causes and what, in turn, causes them. The *coup de grace* is likely delivered to the old climate via new winds and an altered greenhouse, but ocean changes may be what sets up the flip to a new mode of operation - and, even more ultimately, it may be continental drift that sets up the modes into which ocean circulation can shift.

I'll return to this Rube Goldberg chain of causation later, when I fly home over the Gulf Stream up above 60° North, whose warmth encourages the air to rise and thereby helps to stabilize the present cell pattern. But, when the Gulf Stream falters and no longer extravagantly warms the air at 60° North, the atmospheric cells may be vulnerable to disruption by such usual decade-scale climate oscillations as El Niño and the North Atlantic Oscillation.

IF ANYONE LOOKS UP and sees this plane high over the Sahara of northern Africa, he or she is likely to wonder where it is heading. Having heard of a moon landing back in 1969, some might think we were heading there (improbable, but it takes a lot more knowledge to distinguish between the possible and the probable). Others might suppose our destination was their nation's capital, though the more experienced would realize that our 10,000 meter altitude would make that unlikely.

Would the observer watch long enough to judge our direction, which is a little east of south? Would the observer know the map of Africa well enough to realize what lay farther south and east, several countries beyond their immediate neighbors? (A continent with 62 countries, constantly changing their names, would certainly stretch my abilities.) Or have enough education to know about the shortest-and-fastest great circle routes, and that this particular one led to Johannesburg?

So too, our knowledge of gross anatomy attained via butchering grazing animals does not necessarily prepare us for understanding the relationships between the pigs and the antelopes, much less their separate evolutionary paths. Or what evolutionary forces moved things along those paths and not other likely ones. If there are shortest-and-fastest paths, some evolutionary equivalent to a great circle route, we sure don't know about them yet.

SEEING THE SAHEL reminds me that population size always fluctuates. That has a lot of implications for the usual view of evolution, the one that says that improvements in place are always happening when something proves useful. However true that may be, it is slow compared to what happens when population size shrinks and expands. If you want to see how things happen quickly, pay attention to the climate transitions, not the more static periods, and look for things that are amplified by environmental change.

Take the Sahel down below (the "shores of the Sahara" in one metaphor), that transition zone between the arid Sahara and the tropical rain forests. A relatively sparse savanna vegetation of grasses and shrubs now covers most of the Sahel, which stretches from Senegal in the west to the Sudan in the east.

When rainfall improves in the Sahel and the shrunken lakes re-expand, humans move into the newly grassy areas. You don't get a uniform expansion of central population (centered, say, on the Congo) into its neighbors and so filling in the space left by those who moved to the vacant territory. That is simply too slow, compared to the time scale of the climate fluctuations (besides, territories are contested). What you see instead is an expansion of the subpopulation that just happens to live next door when the opportunities open up. In short, the people who are already adapted to living in arid regions get opportunities that the central population doesn't. They're the ones who see the new resources, and when no one is yet contesting them, they're the ones who get more grandchildren as the result.

Now consider the flip side, the Sahel drought that often occurs a few decades after the expansion. There was a severe drought in the 1970s and, if the quarter-century cycle of Sahel rainfall and Atlantic hurricanes holds up, it might again be in trouble. You get hungry people trying to migrate back into the already filled southern regions of the Sahel.

Because the frontier people may have survival skills that are somewhat better than those of settled people, they may do better in the competition. The Huns invading Europe is just a recent episode of an old story. The other thing that happens, of course, is that species explore new foods during the hard times. They are forced to innovate, and some of those new skills may allow them to exploit resources not being used where the central population lives, so the central population density can

For the production of man a different apprenticeship [from forests] was needed to sharpen the wits and quicken the higher manifestations of intellect – a more open veldt country where competition was keener between swiftness and stealth, and where adroitness of thinking played a preponderating role in the preservation of the species.

– RAYMOND DART, 1925

actually increase somewhat when the immigrants arrive. The result is eventually the same: peripherally-useful genes are infused into the more central population.

So far, none of this story is particularly human; the same thing likely happens on a more local scale with baboons, which are monkeys that have adapted to treeless settings but will happily invade woodland. Even some chimpanzees can survive in semi-arid Senegal environments rather than their usual forests (knowing how to dig for roots and tubers can confer survival, as you can get water along with calories). Some anthropologists believe that an important part of our ancestors' story involves learning how to thrive in arid environments. There are a lot of large antelope species that live in such places, and until you learn how to gather grains and bake bread, humans in arid lands likely thrived by being able to eat meat on the hoof.

When we trace food production back to its beginnings, the earliest sites provide another surprise. Far from being modern breadbaskets, they include areas ranking today as somewhat dry or ecologically degraded: Iraq and Iran, Mexico, the Andes, parts of China, and Africa's Sahel zone. Why did food production develop first in these seemingly rather marginal lands, and only later in today's most fertile farmlands and pastures?
– JARED DIAMOND, 1997

It's an interesting "pump the periphery" principle that tends to make useful-on-the-frontier genes much more common in the central population than if the population size were static and only frontier peoples needed frontier genes. Whatever the speed of the apocryphal improvements-in-place, you are far more likely to see substantial changes when climate is fluctuating. That's true for short cycles, like the Sahel cycle, for mid-range cycles like the one every 1,500 years that makes West Africa swing between wet and dry within a generation's time, and for the "glacially slow" ice age changes as well.

And when it's so bad that the central population fragments into isolated refugia, even more dramatic things can happen.

Oceans and deserts are powerful engines in human affairs. The Sahara is another enormous pump, fueled by constant atmospheric changes and global climate shifts. For tens of thousands of years its arid wastes isolated the very first anatomically modern humans from the rest of the world. But some 130,000 years ago, the Sahara received more rainfall than today. The desolate landscape supported shallow lakes and semi-arid grasslands. The desert sucked in human populations from the south, then pushed them out to the north and west. The Saharan pump brought *Homo sapiens sapiens* into Europe and Asia. And from there they had spread all over the globe. . . . But the pump shut down again. As glacial cold descended on northern latitudes, the desert dried up once more, forming a gigantic barrier between tropical Africa and the Mediterranean world. Fifteen thousand years ago, global warming brought renewed rainfall to the Sahara. The pump came to life again. Foragers and then cattle herders flourished on the desert's open plains and by huge shallow lakes, including a greatly enlarged Lake Chad. Then, as the desert dried up after 6000 B.C., the pump closed again, with its last movements pushing its human populations out to the Sahel, where they live today. Like the North Atlantic Oscillation, the Sahara is a pump with the capacity to change human history.

– BRIAN FAGAN, *Floods, Famines, and Emperors: El Niño and the Fate of Civilizations*, 1999

The modern rain forest stretches from the Atlantic to eastern [Congo], which is closer to the Indian Ocean. During glacial maxima, it seems, the rain forest shrank to three small patches, one near each end of its present extent and the third in between, in southern Nigeria and Cameroon. These three oases sometimes added up to no more than about 20 percent of the present extent of the African rain forest.

– STEVEN M. STANLEY, *Children of the Ice Age*, 1996

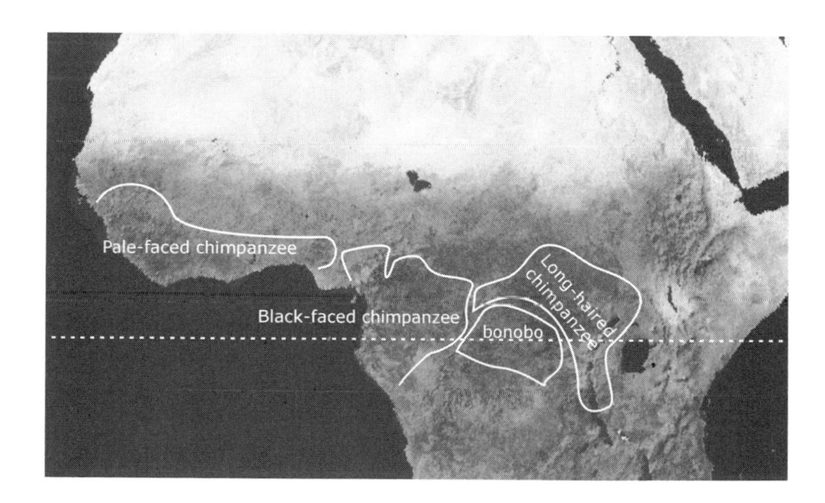

To: Human Evolution E-Seminar
From: William H. Calvin
Location: 0.1°N 21.7°E 10,000m ASL
Latitude Zero

Subject: **Population fluctuations and refugia**

High above the equator gives a view of the right bank and the left bank of the Congo River, the dividing line between our cousins, the chimp and the bonobo, starting about 2.5 million years ago - about the same time that our earliest *Homo* ancestors split off from the bipedal apes, probably somewhere east or south of here. Lots of swamp forest down below, which is where the last bonobos live. (They're called the "Left Bank Chimps" for other reasons, as well.) They were trapped there by climate change, in one of the refugia from the cooling and drying episodes. The bonobos in the other refugia probably died out.

Certainly in the time since the orangutans split off the great ape tree at about 12 million years ago, climate has changed. Globally it cooled - and, in East Africa, the effects were even more pronounced because the highlands of Kenya and Ethiopia were pushed up like a blister, so that present-day Nairobi is slightly more elevated than mile-high Denver. The largest cities of a number of eastern and southern African countries are equally elevated (which is fortunate, as it gets them up above the lower-lying malaria zone).

Temperatures decrease about 6°C for every thousand meter rise, everything else being equal, so Nairobi ought to be about 10°C cooler than Kenya's Indian Ocean coastline - and up here in the stratosphere at 10 km, the air ought to be about 60°C cooler than down on the ground. The two East African blisters thinned out the forest and created more open woodland and even savanna (grasslands with the occasional tree). Something similar probably happened in South Africa, where the first australopithecine skull was found.

At the same time that things cooled, climate became much more variable – so the averages don't tell you the whole story. (Averages are just devices to keep us from thinking more deeply.) Just as variability in populations is the key to thinking about how species can change, so variability in climate (especially the fire-prone droughts) is the key to thinking about what speeds up species change.

WE DON'T KNOW MUCH about how the brain reorganized during the last five million years, though we do know one tantalizing feature of average size. Unlike upright posture which was pretty well established before four million years ago, brain size didn't increase much until the *Homo* lineage was spun off from the australopithecine lineage about 2.4 million years ago. Since then, the brain has increased about three-fold in volume (and more like four-fold in neocortical surface area, from more infolding). Furthermore, sexual dimorphism decreases in the *Homo* lineage; instead of australopith males being twice as large as females (usually a sign of males fighting one another over access to females), the *Homo erectus* males revert to being only 20 percent larger than females, a pattern much like our chimpanzee cousins.

And as I earlier mentioned, the archaeologists say that split-cobble toolmaking also started up about 2.5 million years ago. Why was there all this prehuman action starting about 2.5 million years ago?

The first clue is that it isn't just the hominid lineage that spun off new species about then. That's also when the ancestral *Pan* lineage split into what is today the bonobo and the chimp. The bonobos are confined to the left bank of the Congo River, and now limited to the forest immediately below me, here at Latitude Zero. The various subspecies of the common chimpanzee extend across equatorial Africa from the

Rift Valley into westernmost Africa. About 2.5 million years ago is also about when, among the lesser apes in southeast Asia, the gibbons spun off the siamangs. And numerous other major mammalian groups such as pigs and antelopes also show a lot of speciation happening about then, best documented in Africa. This speciation fest, by itself, suggested big climate change between 3 and 2 million years ago.

The second clue is that Northern Hemisphere glaciation intensified between 3.1 and 2.5 million years ago, thanks to all the moisture delivered to the far north via evaporation from a more vigorous Gulf Stream. Ice sheets eventually built up to the height of mountain ranges over Canada and Scandinavia. The ice mountains tended to melt off every 40,000 to 100,000 years, only to rebuild again. Yet the site of the hominid speciation action was likely in the tropics, probably somewhere here in Africa. And Africa wasn't icy back then - a little cooler (3-5°C), but not the sort of thing that would keep the tropics from being nice and warm in most places. Most importantly, Africa was much drier (Lake Victoria, one of Africa's largest lakes, dried up during the last ice age).

So what do the Ice Ages have to do with stimulating all this evolutionary action among the mammals - and particularly among our African ancestors? (Think drought, not cold.) What might it have to do with the items on my little chunnel-train list of the big-time augmentations of chimpanzeelike behaviors?

First let me explain some of the standard story about climate change and population fluctuations. Then I'll try to show what abrupt climate change adds to the story, which is a much less settled question, not yet in the textbooks.

DARWIN SAW THAT CLIMATE had repeatedly changed but, unlike others before him, he successfully figured out a mechanism whereby animal species could change with it, to adapt body and behavior to the new climate regime. Just spawn a lot of variations in each generation and, given the high mortality among the young, only those variants better adapted to the current environment will survive long enough to reach reproductive age. And those lucky variants will spawn additional variations around their body-and-behavior traits, to further explore "fits" to the environment's opportunities and perils. Those variants better suited to some other climate simply tend not to grow up and reproduce.

But note that this need not be sustainable change. When the climate changes back to the original, the adaptations can track it back again. (Remind me later to explain how speciation can ratchet the adaptation, so it doesn't drift back so easily). Furthermore, adaptations may mostly happen when there is no other choice. At least on some time scales, climate's influence needs to be viewed with some skepticism, as most species react to cooling and drying episodes by moving elsewhere, places where their suite of adaptations still works.

Well, *moving* is something of an euphuism; if there are regional subpopulations, some of them may die out while others continue. With serious climate change, this may leave only a few subpopulations in refugia, places where the species still has all of the essentials for making a living and reproducing. Let climate improve, and they will "expand their range" to live in more places, with refugia pioneers rediscovering those old places where the species once thrived.

Population size is always fluctuating like this. A shrink-and-expand cycle produces more evolutionary change than adaptations-in-place, as I mentioned earlier, but other factors that truly fragment the central population may prove even more important in transforming the species, particularly (as I'll mention in a minute) because of the chance aggregations that occur when things fragment.

Population fragmentations are what happens when a lake almost dries up. As the water level drops, you get a series of small ponds and puddles, in which life continues – but there's now a lot of inbreeding because they are trapped and cannot circulate. There may be some selection for living in the increasingly salty ponds. Most little ponds dry up completely, and the life in them doesn't contribute to what happens later (there are some exceptions, animals whose lay-them-and-leave-them young can survive desiccation). If only the population in one pond survives and then re-expands, we see a classical "population bottleneck" where the re-enlarged population is comprised of only closely-related individuals. Note that much of the pre-existing variety may vanish, even though little natural selection affected the survivors directly (it just eliminated much of their competition by chance). Refugia are common on land, too, and land animals can be similarly restricted to inbreeding for awhile, with a great reduction in genetic variety because of sheer chance. Cheetahs, all very similar genetically, likely re-expanded from one such small surviving population. This means that natural selection no longer has much variation to operate on, preventing evolution until mutations and cross-over breaks eventually generate some new variation on which recombination can act.

But more often, multiple ponds survive the downsizing and fragmentation. When the old lake refills with the rains, multiple small groups form the basis of the re-enlarged animal population. Each group may have survived for a different reason, some developing adaptations but most not. It is presumably only when the land refugia are also under stress, when they too become excessively cool or dry or dusty, that selection can efficiently operate to improve thermal regulation or kidneys or noses. Or to select for rare abilities that, for once, make a difference. This seems fundamentally different from "who survives" during ordinary population contractions into unstressed refugia.

And there is really nothing to suggest that it was all that cold in tropical refugia for our ancestors. Drought, however, is another matter, as is an ecosystem that fire has severely disrupted. Cooling is just the easiest thing to measure in reconstructing paleoclimate, and not necessarily the most relevant thing to survive and thrive.

Lake Victoria, right on the equator over in East Africa, dried up during the last ice age and abruptly refilled about 15,000 years ago. The cichlid fish in the East African lakes split into many new species about then.

Selection during downsizing isn't the only way that evolution operates. There are also opportunities to be exploited when conditions improve. And for an ape-like creature already adapted to making a living on the savanna, the opportunities were of boom-time proportions.

A changing climate drives different populations apart and brings them together again. This could have facilitated speciation.
– RICHARD POTTS, 2001

The later part of the Pleistocene had been a period of extreme fluctuation in climate. Vegetation zones had moved north and south, up-mountain and down. Extensive woodlands had been fragmented by invading steppe or savannas, and had been rejoined as forests returned. Glaciers and harsh periglacial climates made vast areas of northern Eurasia periodically uninhabitable by hominids, presumably spurring major migrations and causing local extinctions. Sea levels had risen and fallen, alternately creating islands and land bridges. Perhaps no period in the history of the globe had been more conducive to the emergence of new species and to competition between related species newly in contact - in other words, to evolutionary change. And the variety seen among later Pleistocene hominid fossils was, in fact, exactly the kind of thing one might expect to find under those conditions.
– IAN TATTERSALL, 1995

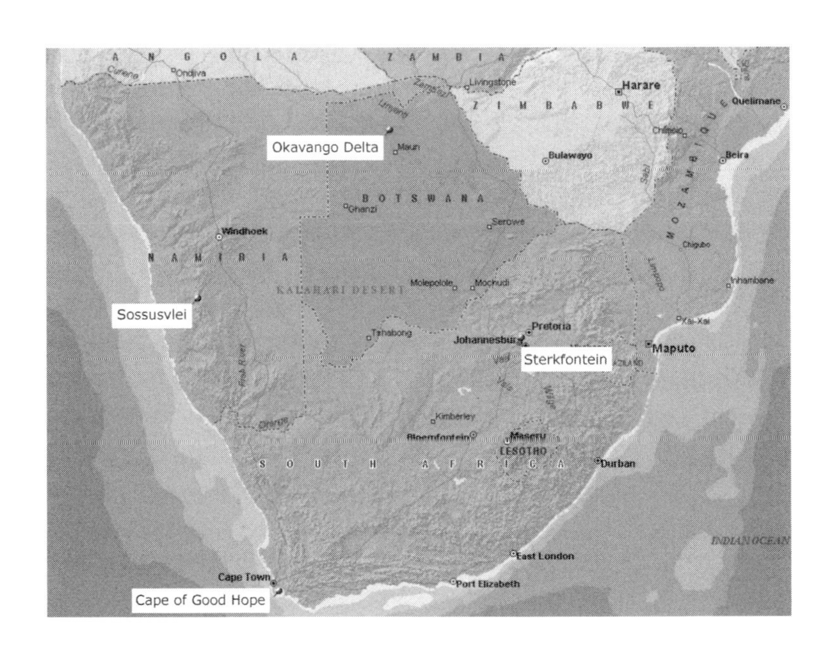
Okavango Delta
Sossusvlei
Sterkfontein
Cape of Good Hope
ANGOLA
ZAMBIA
ZIMBABWE
BOTSWANA
NAMIBIA
KALAHARI DESERT
SOUTH AFRICA
LESOTHO
MOZAMBIQUE
INDIAN OCEAN
Harare
Livingstone
Maun
Bulawayo
Ghanzi
Serowe
Windhoek
Molepolole
Mochudi
Tshabong
Pretoria
Johannesburg
Maputo
Kimberley
Bloemfontein
Maseru
Durban
East London
Port Elizabeth
Cape Town
Quelimane
Beira
Chigubo
Inhambane
Xai-Xai
Ondjiva

To: Human Evolution E-Seminar
From: William H. Calvin
Location: 19.39412°S 22.75876°E 973m ASL
Okavango Delta, Botswana

Subject: **The island advantage**

Here I am in the Kalahari Desert, as far south of the equator as the Sahara is north. It's where the equatorial air that turned south finally comes down from on high, thoroughly dried out. And since the weather systems move across the southern continent from east to west, it makes this mid-continent location a rain shadow as well. The plentiful water hereabouts is rain that fell elsewhere, and then ran downhill to here.

This delta is full of low islands, thanks to hard stuff beneath the shifting sands that the flowing water cannot easily cut into. So when the river comes down out of the mountains of Angola and reaches these sands, it fans out - rather like what happens when a hose is left running in a large shallow sandbox. This cuts a large land area into irregular parcels, looking from above like a reticulated giraffe's skin. (On many such islands, there are even reticulated giraffes nibbling at the tree tops.) At the moment, the water level is low and so a lot of shallow water is now converted into green fields of delicious grass.

Islands are always a matter of interest to the evolutionary biologist. Just as Darwin was the pioneer of the modern theory of evolution, so Alfred Russel Wallace was an equivalent pioneer of island biogeography. Here we see fragmentation without downsizing, a lovely teaching example. Adjacent islands merge when the lake level diminishes, and the islands get rearranged when it floods. Temporarily there may be islands without predators, and others with an oversupply.

And speaking of grasslands-adapted baboons happily invading woodlands, I got a good dose of it last night. They dropped out of an overhanging tree onto the roof of my tented cabin in the middle of the night, shaking things like an earthquake every half hour. At one point three tails could be seen in the moonlight, dangling over the edge of the rain fly.

At least my open-mouthed yawn, when standing at the front door of the tent, was taken as a threat by the baboons, who ran away despite my lack of impressive canine teeth. They didn't even yawn back to display their oversized canines. (I hope no one will argue that human yawning is an adaptation for dispersing baboons. Some people see adaptations everywhere. Yawn.)

I somehow missed the arrival of the three elephants (who tore up the camp's internal electricity cables a few hours before dawn) – I can't imagine how I missed them, as I was awake half the night. Don't these animals ever sleep?

AFTERNOON NOW, AND SO QUIET that the baboons must be indulging in a siesta. The game guide says that when baboons are so active at night, it is because they fear leopards. So I missed seeing a leopard too.

My cousin (no, not the German cousin again; she's the Colorado-England-Kenya cousin with whom I discuss African health planning, her field) came over to where I was writing this morning before breakfast. Just turn around, she said, and notice the elephant on the river bank behind you. Oops. But I could hardly miss the other two elephants in between the cabins, as they were busy dismembering a fallen tree while we ate a proper English breakfast and watched them.

I've just seen two examples of how life shapes geography. The river channels between these islands are often maintained by the hippos when they trudge through any new sandbars. Hippos thereby contribute to maintaining island isolation for other species, preventing them from wading to the neighboring island at lower water levels. Furthermore, the islands themselves are built up to somewhat higher elevations by all the termites that glue the sand grains together into harder stuff. This island is full of tall termite mounds and their subterranean infrastructure.

I ALMOST MISSED NOTICING the baobab tree because I thought, from the distance, that it was just another tall fat termite mound wrapped around a nearby tree, like the one I saw after breakfast. This baobab looks, on closer inspection, like a table decoration, made by standing a potato upright with some toothpick "feet" and then sticking some leafy twigs up near its top, to mimic a broad-brimmed hat. Baobabs are thought to be very important in hominid gathering strategies, as its leaves, fruit, and seeds can all be used (its pollen even makes a good glue). And it can provide water.

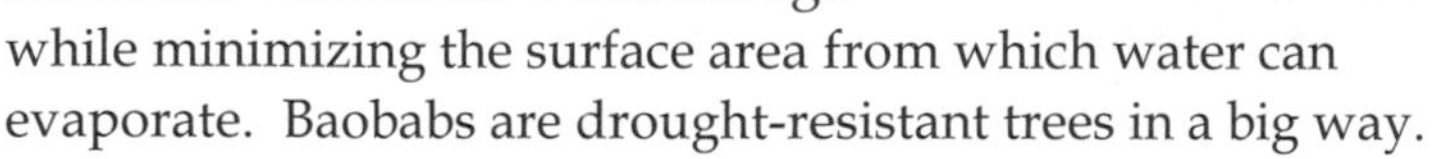

Baobabs are another little lesson in surface-to-volume ratios, and how you maximize volume for water storage while minimizing the surface area from which water can evaporate. Baobabs are drought-resistant trees in a big way.

Humans (especially marathon runners) sometimes do the opposite, like the high surface-to-volume ratio of the tall skinny Maasai compared to the rest of us. It is said to be an adaptation for losing heat quickly by evaporation, by maximizing the surface area from which to sweat. You have to avoid cooking that big brain with the heat from running and from the hot sun, particularly from both at once. Upright posture itself is a way of minimizing surface area exposed to the hot overhead sun, just head and shoulders taking the full hit rather than a broad back. You minimize your shadow. Some anthropologists suggest that upright posture is a savanna adaptation for treeless places where you can't "shade up," as most sensible animals do at midday.

There are so many suggested explanations for upright posture - the evolutionary biologist J. B. S. Haldane liked to observe that only a human can swim a mile, run 20 miles, and then climb a tree - that it is difficult to say why our ancestors did it and why the other great apes didn't. Sustained bipedal running is really more efficient than quadrapedal, especially if you have a heavy head to support (cantilevering it during bouncy locomotion sure takes a lot of neck muscles, compared to balancing it atop the spinal column). This is an advantage in the long run, which must be distinguished from arguments about how bipedality got started.

I have a favorite, naturally, but it doesn't really exclude any of the other candidates for bipedality's origin. Most leading features in evolution have a supporting cast. Indeed it is often like a repertory theater, with the star one night acting as the walk-on butler the next night. Picking an overall "star" in an evolutionary repertory is often a mere matter of taste, though "which is fastest" is a good criterion when you are trying to figure out how we got here from back there, and so quickly.

AS I MENTIONED IN PARIS, the chimpanzeelike hips and knees got modified early on, presumably for upright locomotion - but there are nonlocomotion possibilities too, such as upright stance per se. Seeing over the tall grass was said to be an advantage (until they discovered upright posture came four million years earlier than life on the savanna). Upright stance is also an advantage if you wade a lot but haven't yet learned to swim.

There may be some secondary uses of upright stance, as for picking fruits off trees without having to climb them, but the others seem more likely to have substantial payoffs. There are also some temporary advantages of upright posture for hunting large animals, as naïve animals tend not to fear them as they do their usual four-legged predators. This allows the hunter to get closer before they start edging back. But this advantage doesn't last, once a herd has been hunted for awhile.

Most animals that live in the grass and bush have, of course, managed to do it without switching to upright posture, so I tend to favor the wading-and-shallow-diving hypothesis, given the suite of other adaptations that we humans have (subcutaneous fat layer, copious tearing, loss of most body hair, breathing control for underwater, kidneys that are relatively unconcerned with hoarding salt and water, and so forth) that are often seen in the mammals that returned to the sea. Losing hair for whatever purpose would also tend to promote upright stance because infants would then need to be carried, being unable to find much maternal hair to grasp (we are an exception to the general infant-carrying rule among primates). Or perhaps the infants lost their ability to cling, forcing carrying.

This is usually called the "aquatic ape hypothesis" because it involves so many things that are also seen in the land mammals who returned to life in the sea (whales did it 100 million years ago, sea otters only several million years ago) with a salty diet and a need for less body hair. No one imagines a fully aquatic ancestor, so the name is somewhat misleading, but rather a creature that foraged along shorelines and occasionally swam a little. One version of this hypothesis about hominids emphasizes islands, as they have a lot of shoreline and, in a drought, all of the remaining resources on the island might have been the fish and shellfish along the shoreline.

Parts of [the world] are neither land nor sea and so everything is moving from one element to another, wearing uneasily the queer transitional bodies that life adopts in such places. Fish, some of them, come out and breathe air and sit about watching you. Plants take to eating insects, mammals go back to the water and grow elongate like fish, crabs climb trees. Nothing stays put where it began because everything is constantly climbing in, or climbing out, of its unstable environment.

– LOREN EISELEY,
The Night Country, 1971

While apes isolated on a chain of islands in the Red Sea would indeed be an excellent setup for doing the aquatic adaptations quickly, it may be that the lakes and rivers of East Africa could have also done the job. There are lots of fish in shallow lakes that can be herded into modern nets by small boys splashing around. In the days before nets, one could likely drive them into restricted spaces where someone could heave them ashore.

Even filling in the fossil gap between the great apes and the australopithecines may not help settle the issue because they will yield mostly bones from the usual sites of preservation (caves, lake margins). Bones often tell you something about muscle strength and, via the size of vertebral openings for nerves in the chest region, something about how good their breathing control might have been. But they won't tell you about fat layers and the extent of hair coverage, nor about how much salt was conserved by their kidneys.

You might think that, because so many hominid fossils are found at former lake edges, this might be used as evidence for shoreline foraging. But the pros all know about preservation bias - that sites like forests are very unlikely to preserve bones, and that lakeshores are excellent in that regard. Caves also preserve the occasional skeleton as at Sterkfontein, but no one assumes australopithecines preferred to live there (more often, pits within Sterkfontein became death traps for explorers or those being chased). And so paleoanthropologists quite understandably treat the water's edge as simply a great setup for preservation (a flooding lake buries those who die near the water's edge and moves the shoreline - and the grazing animals who might crush the bones - back away), and do not also use it as an argument for where they preferred to live. Still, the evidence is often consistent with shoreline living.

In forty years, the aquatic aspect has gained few adherents among the pros, even though the savanna bipedality hypothesis has recently proven awkward for the australopithecines. Archaeologists do not buy the aquatic hypothesis, perhaps because there's nothing in it for them to study yet (a nice trash heap of shells that is 5 million years old might change their minds). Physical anthropologists don't like it for similar reasons; their strength is anatomy, and most aspects of the aquatic ape hypothesis are physiological. When they proclaim "But there is no evidence for that!" about the aquatic hypothesis (and they are quite vehement), what they seem to mean is that there is none of their specialized kind of evidence - except, of course, those hip and knee rearrangements, and they prefer to ascribe other functions to them. Everybody has a mental checklist for what needs explaining (mine is that chunnel-train list), mostly items within their own technical expertise, and many don't like to be bothered by things that don't address their agenda.

Occasionally in the history of science, facts finally accumulate to the point that the old way of looking at them seems a little awkward compared to another - perhaps a minority view or some new suggestion. The facts aren't yet good enough to make the woodland-to-savanna bipedality hypothesis look awkward without a shoreline interlude. But then the suite of hominid questions that require evolutionary answers doesn't, for most, yet include the physiological or the neurobiological agenda.

Now that the evidence for upright posture has reached back to six million years ago, very close to the DNA dating for the common ancestors with chimps, we are faced with a situation where efficient bipedal locomotion (losing those tree-climbing feet to improve running) happens a few million years after upright stance per se. So maybe infant carrying or something like wading came very early. Certainly the bipedal apes were sticking close to woodlands and even *Homo erectus*, though found in more arid environments and adapted to heat stress, probably had the same savanna drawbacks that we moderns do: our kidneys waste so much (compared to truly arid-adapted animals) that, before canteens were invented, our ancestors had to stay close to drinking water.

WE NOW KNOW a lot about island biogeography, including the fact that evolution seems to operate faster in small populations than on continents with large ones. Large central populations tend to buffer change, as natural selection for one trait may be diluted or balanced out by selection for another. Individuals there have a lot of mating choices, and aren't as likely to mate with someone whose ancestors have been through similar selection regimes.

But the archaeologists are now starting to emphasize that population density of australopithecine and *Homo* species may have remained quite low throughout the ice ages, meaning that large central populations may merely be a feature of recent agricultural times. Low numbers are what you would expect from top predators in the food chain, the same reason why bird-eating birds like peregrine falcons are so few in numbers compared to pigeons, or why it takes large herds of grazing animals to support a few lions. Or a few hominids. Maybe it wasn't until we learned to grow grains and bake bread that human population density could increase significantly. Still, I'd bet that hunting is what most allowed the hominid range to expand, what with all those naïve herds to tackle.

HAPPENSTANCE DURING SUBDIVISION may omit typical predators. For example, in the last warm period when rising sea level converted a peninsula on the coastline of France into the island of Jersey. The red deer trapped there underwent a considerable dwarfing in stature within only a few thousand years. That's probably because their usual predators died out locally – predators that had made large body size a real advantage. Lacking predators, there is something to be said for maturing early (at a small body size) and having more time to produce more offspring.

So it's possible to predict some of the things that might happen if a prehuman population were fragmented into smaller inbreeding subpopulations by an abrupt climate change. A higher percentage of the total then live on the margins of some habitat (it's surface-to-volume ratio, once again) and the margin is also where selection pressure is greatest.

Local extinctions, as when an island population becomes too small to sustain itself, also speed evolution in a way that isn't immediately obvious - that's because *no* competition is markedly better than *some* competition. A local extinction creates an empty niche. When subsequent pioneers rediscover the unused resources, their descendants go through a series of generations where there is more than enough food. That means that even the more extreme variations that arise, the ones that in childhood would ordinarily lose out in the competition with the more optimally endowed - such as the survivors of a resident population - can now survive and reproduce for a few generations.

When the environment again changes, some of those more extreme variants may be able to better cope with the third environment - better, at least, than the narrower range of variants that would reach reproductive age under the regime of a long-contested niche. So a flipping climate has an ability to get more variants out onto the board in play, as well as providing a recurring stress that culls the less versatile.

Thinking of the Ice Ages as the "Chattering Ages" with alternating boom-and-bust provides a perspective quite different from adaptationism's usual focus on efficiency. Efficiency arguments, as I mentioned, tend to suggest lean mean machines without a scrap of excess baggage. But the need to discover a new way of making a living within a single generation shows how jack-of-all-trades variants could survive better. Techniques that were last needed a hundred generations earlier would need to be rediscovered, in order to make use of less-favored food resources.

RANDOMLY-PICKED SMALL POPULATIONS are rarely average. Often they have some odd clustering. This always seems to surprise people - as, say, when a randomly-selected jury turns out to be all men or all women, not exactly the proportions in the larger population from which they were drawn.

This usually isn't a bias in the selection procedure; it's just how chance sometimes operates when a few are drawn from the many. This happenstance clustering has some interesting implications for the evolution of social traits, things like language or reciprocal altruism where groups are important. By chance, some subgroups are strikingly overrepresented in one trait, woefully lacking in others. Evolution now operates on dozens of subpopulations independently, rather than upon the whole large "average" population - and a subpopulation may thrive relative to others, simply because it chanced to have a disproportionate number of the bearers of some minority trait.

Culture can pass things along, but a critical mass is sometimes needed to get cultural transmission going; most inventions simply die out. Others are useful in the long run, but can be easily overburdened in the short run - and that's the big problem with reciprocal altruism.

IN MOST SPECIES THAT SHARE FOOD more generally than just mothers sharing with their offspring, individuals only share with close relatives. If they help out someone being attacked, this assistance is also usually limited to close relatives. That's kin selection, where you are helping out copies of your own genes by helping the others.

As human society presently demonstrates, there are great benefits to expanding the circle of beneficiaries to nonrelatives, what is called reciprocal altruism. But it's a puzzle: How could that happen, when everyone loves to freeload? Cheaters (those who receive without eventually reciprocating) are the norm in animal societies. Any individual that tended to give away food, or indiscriminately risk life and limb for non-relatives, would be a loser - unless living, by happenstance, in a subpopulation with a lot of other indiscriminate sharers, likely to provide benefits at other times.

And that's what the repeated fragmentation of large prehuman populations into many smaller subpopulations could have occasionally created: a group with a critical mass of sharers. In hard times, when the every-man-for-himself groups were wasting a lot of time and effort at fighting over the remaining food, the groups that shared (and otherwise minimized conflict) might have survived better, successfully raising a next generation when the others were squabbling. They weren't competing against each other as in team sports but rather against the downsizing environment, for sheer survival.

In this manner, natural selection can occasionally operate on groups – and therefore on social traits. Some things, like language and altruism, only operate between a substantial number of individuals. If all the subpopulations are lumped together and mixed, as in today's cosmopolitan societies, it may be hard to initially evolve such traits, simply because there are always enough freeloaders nearby to swamp and sink even a promising startup. But a history featuring fragmentation, and then amalgamation (and repeating hundreds of times), is capable of accomplishing some things that might otherwise be improbable. Thinking in terms of the average can seriously mislead you.

In our African idiom, we say, "A person is a person through other persons." None of us comes into the world fully formed. We would not know how to think, or walk or speak, or behave as human beings unless we learned it from other human beings. We need other human beings in order to be human. The solitary, isolated human being is really a contradiction in terms.

– Archbishop Desmond Tutu, 2000

The traditional thinking that dismisses group selection is that, even if a subpopulation happened to have a majority of cooperators, you'd still expect that tendencies to share could be swamped by all the non-reciprocating freeloaders, who would out-reproduce the sharers and slowly sink the altruistic practice. So the group trait would be leaky, like a car tire going slowly flat.

If this were the prime consideration, of course, we would also have to conclude that car tires would never work. Sooner or later, they too all go flat. We just pump them back up occasionally, and maybe that's what reciprocal altruism takes. The bust-then-boom cycle provides both a concentration mechanism (via fragmentation) and a pump (survivors get the eventual re-expansion opportunities). Such pumping might allow widespread cooperation to become established long enough for other things to be invented that prop up cooperation by combating freeloading. I sometimes think that the first sentence spoken was "But you owe me!"

If you could interview a chimpanzee about the differences between humans and apes . . . , I think it might say, "You humans are very odd; when you get food, instead of eating it promptly like any sensible ape, you haul it off and share it with others."

– GLYNN ISAAC (1937-1985)

To: Human Evolution E-Seminar
From: William H. Calvin
Location: 24.65469°S 15.90924°E 858m ASL
Sossusvlei, Namibia

Subject: **Hominid opportunities in deserts?**

We've gone from part of the Kalahari Desert, the part with a lot of water passing through and evaporating in Okavango Delta, to a more typical desert farther to the west, one with ephemeral streams lined with a few trees – and almost none elsewhere. It's a savanna strip.

Admittedly, this was a good year for rainfall, but even when it is much drier, the oryx, eland, and springbok (you really have to see the youngsters' jack-in-a-box act to appreciate the name) thrive here on the valley floors among the sand dunes. When you fly low over the area, you see a series of dark lines stretching across the light desert floor. They are dry watercourses. There is still some water beneath the surface and, even if there isn't, the nearby plants are good at storing water for the rest of the year. The leaves of some plants can be crushed and wrung out like a washcloth, yielding a surprising amount of water.

When the South African paleontologist Elisabeth Vrba told us that there were a lot of new species of antelope appearing back about 2.7 million years ago in southern Africa, one of its prime implications was that it indicated a need to adapt to arid environments. And today, there are amazing numbers of oryx and the even bigger eland here in the Namib Desert, happily grazing. There are some small antelope that can forego visits to the waterhole entirely, getting enough water from the leaves they eat. If we could test antelope for landscape esthetics, they'd probably prefer to look at something like a bush airstrip – a cleared area where predators can't hide. And so bush pilots coming in for a landing have to first buzz the airstrip, to chase them away.

It makes me realize how much meat on the hoof there was for our ancestors to have exploited in arid environments. If the antelope and the desert elephant hereabouts can adapt to the desert, then they may make it possible for their predators to do so as well. They just have to patrol the strips, like teenagers cruising those urban strips following highways out into the countryside. All the resources are along a track, and it's just back and forth.

The fauna associated with the small-brained australopithecine fossils indicate a wooded environment; they wouldn't have liked it here. The later versions called *Paranthropus* between 2 and 1 million years ago were sometimes found in wetland environments, as were the earliest *Homo* species. It's *Homo ergaster/erectus* and later species that are found in extremely arid and open landscapes like this. It seems pretty clear what they were eating; certainly in South African coast archaeological sites, there are a lot of eland bones.

THIS PLACE IS A DESERT because, at these southern latitudes, the rains come from the east. And by the time that they have traversed the whole width of Africa from east to west (we're just inland from the west coast), it is even more of a rain shadow than Botswana's Kalahari Desert. Namibia is one of the driest places on Earth. Indeed, what moisture Namibia gets often comes from dew, thanks to the fog drifting in off the ocean, much the same as in the coastal regions of Peru on

South America's west coast. And there is fog here because the cold Benguela current offshore causes the sea breezes to drop below the dew point as they blow in over the cold current.

This cold current is part of the developing story about how ocean currents and winds rearranged themselves just before the ice ages started. Strong winds often sweep surface waters aside – I'll get into the subject when I fly home over the North Atlantic and discuss the conveyor belt for salt and heat – and bring deeper waters to the surface. Deeper waters are cold; they're also loaded with nutrients, and so when they get brought up near the surface where sunlight can penetrate, they serve as fertilizer for all the sea life offshore, a whole food chain worth (lots of seals and dolphins and fishermen hereabouts).

Well, at drilling site 1084, less than an hour's flying time west of here, the surface waters are about 10°C colder now than they were 3.2 million years ago. That means some stronger winds have developed in this part of the South Atlantic Ocean. The big changes were between 3.2 and 2.1 million years ago, in the second half of the Pliocene, just before we start talking about the Pleistocene's ice ages. They go in lockstep with the changes in the North Atlantic. It's all part of the story about how the ocean and atmospheric circulation rearranged themselves, to plunge us into the fickle climates of the ice ages. Which we are still in.

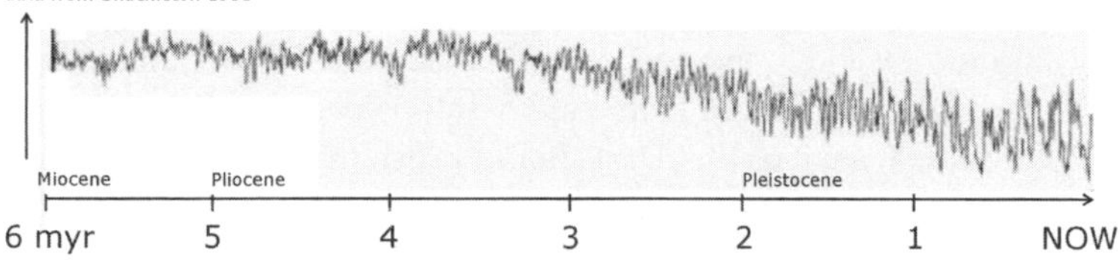

JUST TO PREPARE YOU for hominid fossil country (my next stop is South Africa's caves, then Kenya's Rift Valley), let me suggest reading an account of how hominid fossils have been found. I'm particularly fond of Alan Walker and Pat Shipman's *The Wisdom of the Bones* which contains the following account of how the first australopithecine came to scientific attention in 1924 and how it was mostly ignored for the following decades. I always tend to think of the Leakeys' 1959 discovery of Zinj at Olduvai Gorge as the first hominid skull discovery, but it was only the first in East Africa (and in a stone-tool context). The Taung skull was actually discovered a quarter-century earlier by workers in a South African quarry, but dismissed by experts as some sort of ape:

> It is a classic story of anthropology, all the more engaging for being true. With unerring timing, the box with the missing link in it turned up as [Raymond] Dart was dressing, in wing collar and morning dress, to serve as best man and host for a friend's wedding. The men from South African Railways who staggered up to the house that summer day in 1924 left two large crates blocking the stoop shortly before the guests were to arrive. Dart had them moved to the pergola, where they would be out of the way, and left off dressing to find a crowbar to pry them open. The contents of the first box were uninteresting scraps of fossil eggshells and turtle scutes (the bony plates that underlie the turtle's shell). On the top of the rubble that filled the second crate, Dart spied an extraordinary thing: a natural, fossilized cast of the brain – an endocast. It was of some creature whose brain was about as big as that of an adult chimpanzee. From his work with Elliot Smith, Dart immediately recognized that this was no ape endocast (unparalleled as that would have been), but one with distinctly human anatomy. He rummaged through the box frantically and found a piece of bone, covered in rock, into which the endocast fit. And then real life intervened. The groom appeared, anxious that Dart should brush the dust off his suit and struggle into his stiff collar; the wedding party was arriving momentarily. Dart took these two precious pieces of our ancestry and locked them in his wardrobe, reluctantly abandoning them until the festivities were over. . . .
>
> Dart's precious find was not only overshadowed [first by the Piltdown hoax and then by Peking Man], it was literally abandoned – left, in its humble brown-paper-covered box, in the

backseat of a London taxi by Dart's wife. It was recovered only after frantic searching. Dart gave up on plans to publish a monograph and returned home, discouraged and defeated. He gave up fossil work for many years and subsequently suffered a nervous breakdown. For years, the Taung child sat, forgotten, on Dart's colleague Gerrit Schepers's desk at Wits. . . .

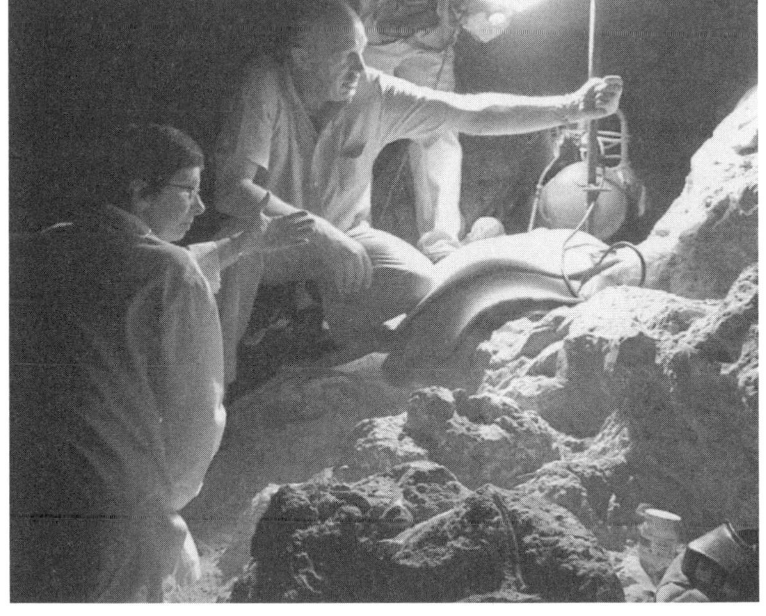

To: Human Evolution E-Seminar
From: William H. Calvin
Location: 26.01552°S 27.73381°E 1,480m ASL
Sterkfontein Caves, South Africa

Subject: **The big change in hominid diet**

There are a series of hominid fossil sites in this beautiful valley outside Johannesburg, now a World Heritage Site. The most famous is the Sterkfontein Grotto, where the "Mrs. Ples" australopithecine skull and endocast was found. Most of the extracted fossils are now in a walk-in vault at the Wits medical school in Johannesburg, where my wife and I spent the day.

It's hard to say which was more impressive: Phillip Tobias showing us the original fossil skulls and comparing them feature for feature with chimpanzee skulls, or Ronald Clarke and Kathleen Kuman showing us an ancient skeleton still in the rock at Sterkfontein's Silberberg Grotto.

In the vault, incredible variety and hard-earned detail. That natural endocast of the Taung child has amazing imprints of blood vessels and the rounded hills and valleys on the brain's surface.

At Sterkfontein itself, first navigating a dizzying aerial skywalk over the surface excavations, ducking under the wire grids that hang over the area to provide a coordinate system. And then scrambling into a dark cave, though several grottos, and eventually coming to a handheld spotlight – showing several figures huddled over a skeleton emerging from the hard rock.

There is likely a complete australopithecine skeleton embedded in the rock at Sterkfontein. It's been there for about 3.33 million years, judging from the other animal species found nearby. As I noted earlier, most surviving skeletons from East Africa were buried in mud, as a lake expanded and covered them up (or they drowned offshore, failed swimmers). This kept some of the bones from being trampled into fragments by herds coming to drink. But the Sterkfontein skeletons were instead embedded in very hard stuff, what makes stalagmites and stalactites. They are emerging very slowly as Ron Clarke uses the miniature dental version of a jackhammer to remove the surrounding brecca grain by grain.

The foot of *Stw 573* has already told a big story. While "Little Foot" is manifestly adapted for upright stance and stride, with a heel more capable of weight-bearing than a chimp's, the big toe is almost as mobile as that of a chimpanzee. So, despite Little Foot's upright adaptations, the big toe might have been of more help in climbing trees than a modern one. The lack of transitional forms used to be a big objection to Darwin, but with a lot of hard work, scientists are finding important transitional forms.

The transition to upright posture is still an unsettled issue in paleoanthropology, as I mentioned when discussing the island advantage. Upright posture is likely all tied up with the transition to gathering and hunting (as opposed to the chimp's eat-it-on-the-spot and snatch-and-grab). There are a lot of aspects to gathering (eggs, seeds, nuts, leaves, roots – and the selection is greatly enhanced by food preparation from soaking and pounding through to cooking). Ditto for hunting (grabbing the defenseless young hiding in the tall grass, surrounding in the chimpanzee manner, maneuver via stampeding over a cliff, and projectile predation of many sorts).

HUNTING IS ONCE AGAIN part of the overall explanation for our ancestral way of life, what with the evidence for eating a lot of grass by 2 million years ago. Grass, not leaves or fruit. We weren't baking bread back then (which is the way we now ingest grass directly). No, some other animal probably converted the grass into meat on the hoof, and the hominids got the characteristic carbon stable isotope ratio from eating the grazing animal – they ate grass indirectly, at one remove. Had they eaten monkeys who in turn had eaten fruit, the bones would have the nongrass carbon isotope ratio instead.

It used to be that you couldn't mention hunting without someone undertaking to correct you. (Phillip Tobias isn't correcting me; he's just demonstrating another use of a hominid femur.) If it wasn't Lewis Binford's minimalist approach to the archaeological evidence back before cut marks were found overlaying carnivore tooth marks on bones in 1979, it was someone saying that gathering was far more important, that even in surviving nonagricultural tribes, meat is only a small percentage of the calories. Ditto for the chimpanzees which hunt.

All quite true, so far as I know, though the Inuit are obviously an exception since they have very little to gather for so much of the year. Indeed, gathering as more important had a certain attractive logic to it. It seemed to follow from the economistlike view of evolution, which is a common misconception about the nature of evolutionary change. So perhaps I should undertake to explain why everyday importance is not necessarily what you want to focus on. This will be heresy to some, old hat to others, but here goes.

DARWIN OVERSOLD GRADUALISM to some extent, in his effort to show that there were ways other than catastrophic change for evolution to occur (recall that nice algorithmic crank that I mentioned back at Down House). This is an Alice-in-Wonderland sort of principle (the Red Queen told Alice that you have to keep running just to stay in the same place). Certainly "automatic gradualism" is the lesson that people most easily remember a week later, rather than Darwin's nuanced notions of evolution.

But there is really nothing automatic about evolutionary change, where everyday usefulness is automatically rewarded (and so you can remember a single principle rather than all those messy details). Biological adaptations can backslide, if not backstopped by speciation, just as cultural improvements are notoriously easy to lose (recall how the Tasmanians lost fire-starting and fishing techniques, once isolated from Australia).

More important, to my mind, is that many adaptations and inventions don't have growth curves. If you invent a digging stick to expand the range of gathering possibilities, there isn't too much you can do to improve it into a shovel until a lot of ancillary improvements occur in the creation of sharp tools of the sort needed to make other tools. To redouble your payoff in terms of calories gathered via the digging stick takes a lot of further invention. If you invent soaking to help remove the bitter taste from plant toxins, it is again hard to double and redouble your payoff until you invent boiling. Efficiency improvements are often difficult. The long stasis in rock toolmaking doesn't surprise me a bit.

The same is true of most aspects of hunting. But one aspect of hunting, projectile predation, has an extraordinarily long growth curve because the distance achieved with accuracy can be redoubled so many times. No matter how many times you've improved your throwing distance, doubling again has additional payoffs in terms of how many days each week that your family gets high-calorie, low-toxicity meat. Furthermore, there may not be much to gather in certain commonplace climate transitions, and getting through the crunch and its aftermath may turn out to be far more important than everyday efficiencies.

Hominid hunting (beyond the chimp and baboon opportunistic style) is now thought to go back several million years. You can tell a lot from how much chewing had to be done. Gorillas have to chew fifty pounds of plant foods every day, and their skulls have a lot of extra space to attach the muscles, all those ridgelines down the middle and around the back. But in the *Homo* skulls, all that anchoring space starts to disappear, as if chewing was no longer such a problem. The cross-section of the jaw muscle can be gauged from the space inside the zygomatic arch that goes from face to ear – I can fit six fingers into the muscle's space on my *erectus* skull cast, but barely two into the same space on a modern *sapiens* skull. Notice too the decreasing size of the teeth.

So, as was obvious even back in the gathering-is-politically-correct days, there has been a big change in diet since the time of the common ancestor with the chimpanzees and bonobos. While cooking – which has made life safe for vegetarians, because of a major expansion in what plants can become food – is part of that story, it may arrive late in the story, long after the major bony changes are seen (although there is one important proposal that would place it back at 1.88

million years when *Homo erectus* arrives on the scene). Gorillas are vegetarians and it has trapped them in a limited ecological niche that requires fifty pounds of rough food every day, with gut length and jaw muscles to match.

There are strong signals of a massive improvement in dietary quality between *Australopithecus habilis* and *Homo erectus*. These include the 60 percent increase in female body mass, a large reduction in tooth and jaw size, and . . . a substantial reduction in the size of the gut. Additionally, this period [about 1.88 million years ago] witnessed a continuing increase in brain volume beyond the size of the living African apes. *Homo* seems to have eaten better than *Australopithecus*. Why is that?
– RICHARD WRANGHAM, 2001

One food source, occasionally exploited by chimpanzees in the more arid areas, is underground: the underground storage organs such as tubers and other roots. They occur only infrequently in rain forests but are found more often in woodland on the fringes of forests, where herbs benefit from storing water and nutrients during dry periods. Most animals can't eat them, as they are rich in toxins and hard to dig up. Mole-rats and pigs specialize in them, and their bones are found in association even with 4 million-year-old hominid fossils. Pigs are particularly interesting because their molar teeth (whose size and enamel thickness tell you about diet) have a lot of similarities to those of the australopiths.

At some point (the problem is when, early or late), our ancestors learned how to exploit such roots as a fallback food when their preferred foods were in short supply. To extract them often requires digging and chopping. They are improved by pounding and soaking, which involves a degree of food preparation not usually seen in the other apes. Cooking is even better. One formulation has it that eating grazing animals goes back to 2.5 million years (there are butchery marks on the bones of large mammals, back about then), and that cooking (or some other equivalently major improvement in food quality) starts about 1.9 million years ago.

GROWTH CURVES are where there is a payoff to repeated improvements. If some brain enlargement is good (we usually assume), perhaps a little more is even better. Curves sometimes turn over, of course, why self-medication based on more-is-better can be dangerous. Physicians since Hippocrates have been warning about the trouble that the more-is-better metaphor gets us into.

But upward growth curves are especially relevant when you have something like abrupt climate change that can pump you up the curve. When there isn't much to gather, hunting temporarily becomes rather important, even if providing less than 10 percent of calories in other times - and one aspect of hunting, accurate throwing, has a nice long growth curve.

Should civilized man ever reach these distant lands, and bring moral, intellectual, and physical light into the recesses of these virgin forests, we may be sure that he will so disturb the nicely-balanced relations of organic and inorganic nature as to cause the disappearance, and finally the extinction, of these very beings whose wonderful structure and beauty he alone is fitted to appreciate and enjoy. This consideration must surely tell us that all living things were not made for man.

– ALFRED RUSSEL WALLACE, *The Malay Archipelago,* 1869

To: Human Evolution E-Seminar
From: William H. Calvin
Location: 34.35729°S 18.47400°E 1m ASL
Cape of Good Hope

Subject: **The turning point that wasn't**

Speaking of pumping, tides must be the original stroke-after-stroke device. Tidal fish traps show you how our ancestors could have gone fishing along this coastline, even without spears or fish hooks. If the tide flushes into a gully, all they had to do is to pile up some boulders to half the height of high tide, making a sill across the gully entrance. It doesn't have to hold back the water effectively, only the larger fish that aren't quick enough to notice the falling tide. Now the pump has a fallback. And so some fish will remain behind the dam, high and drying, there for the taking on your daily visit.

While the evidence for such things along the southern coastline probably doesn't go back more than several thousand years to when the pastoralists arrived, this is the sort of thing that even bipedal apes might have been able to do. There are many natural examples, called tide pools, that trap small fish and crabs every day. These could have served as examples and indeed you can block up the low points of natural tide pools to enlarge them, giving you the idea for how to approach the problem on the larger scale of a little inlet. One can easily imagine people piling up rocks to fill in the low points of the entrance. Groups would have zealously guarded the good gullies, perhaps even settling down for a while.

The cold current keeps the ocean here about 11°C, in contrast to the 20°C waters east of Cape Agulhas. And so the modern version of the fish trap is the tidal swimming pool, sometimes seen in the beaches of the western Cape Town suburbs. A particularly high tide flushes out the shallow pool with new water every two weeks. The sun then warms up the trapped water enough for you to swim comfortably. And maybe you'll encounter a fish, too, looking for a way out. It's too bad that there isn't some way for the tides to pump the water higher and higher with each tidal extreme, with a ratchet tracking it up. You have to wait for biological evolution before such ratchets happen.

There are a series of caves facing out onto the Indian Ocean east of here, such as Nelson's Bay Cave and Klasies River. Klasies is particularly important, as it has remains of anatomically-modern (except for chins) *Homo sapiens* from before 100,000 years ago, back into the last major warm period. Its tools are the usual Middle Stone Age types, and the bones include a lot of eland and penguins, together with some Cape buffalo and wild pigs. Nelson's Bay Cave, a deep cave somewhat above the high-water line which we explored several days ago, has layers from the last ice age and into our present warm period; their tools included projectile weapons, and, judging from the bones they discarded, they had gotten good at bringing home the bacon and dealing with the ill-tempered Cape buffs, among the most dangerous animals in Africa.

THIS SOUTHWEST CORNER of Africa is about as far south of the equator as Los Angeles is to the north of the equator. However, L.A. doesn't have ostrich wandering the bush or shoreline-patrolling baboons snatching your unguarded picnic basket. Indeed, this place looks like a high mountain plateau, treeless with heathers and high winds. You expect to feel breathless due to the thin air but it is located at sea level.

The lighthouse is on another point just a kilometer east, Cape Point, which is much higher and looks down on the Cape of Good Hope. The ocean stretches west to Argentina and south to Antarctica. Bartholomeu Dias reached here in 1488 in the intrepid Portuguese caravels inspired by Prince Henry the Navigator. John II of Portugal named the cape *Cabo da Bõa*

Esperança, Portuguese for Cape of Good Hope (the "hope" refers to trade prospects with China), when his explorers returned and presented their map.

That the Cape of Good Hope wasn't named something else sixty years earlier by a Chinese admiral, Zheng He, is one of the major ironies of history. The Cape is not, by a half degree, the southernmost extent of Africa (that's Cape Agulhas farther east, where the ocean's name officially changes from Indian to Atlantic. Agulhas would not have impressed a sailor very much, being just another protrusion of a wandering shoreline, important only in retrospect while tabulating latitudes.

But when you round the Cape of Good Hope going west, the water turns cold, kelp beds line the shore, and the sea coast stretches away to the north, seemingly forever. You know that you've turned a corner. The Chinese sailors would have wondered where it led, and might have gone on to discover Europe. But it appears that they didn't make it quite this far.

Between 1405 and 1433, so-called "treasure fleets" sailed from China to explore India and the East African coast. The scale of the venture, hundreds of ships at a time, makes European exploration pale by comparison. The Portuguese, later that same century, were far less ambitious than the Chinese but they succeeded. Three ships, led by Vasco da Gama, eventually rounded this cape in 1497 and then continued on to India in 1498.

So why didn't the Chinese discover Europe, and beat the Europeans to the Americas and Australia? Politics. China's explorations were suspended before their ships ever turned the corner into the Atlantic Ocean. Back home, power shifted to a traditionalist faction that wanted to distance China from the outside world. The eunuchs in the government who supported exploration fell into disfavor. By a century later, the act of venturing abroad in a multimasted ship, even to trade, had become a criminal offense in China. They even abolished mechanical clocks after leading the world in clock construction technology.

For an equivalent, you have to imagine an ultraconservative takeover of the U.S. that, for some reason, frowned on both airplanes and computers, with Seattle and Silicon Valley in deep depression and the conservative radio talk show hosts beginning to look liberal in comparison to the people in power. The short-sighted in China set the stage for Europe's 400-year domination of coastal Asia, when it might have been vice versa.

As Jared Diamond is fond of pointing out, various European countries did equally stupid things but, since Europe was never politically unified in the way that China was, there was always some European country to keep exploring when others stagnated. As the story of Christopher Columbus shows, sometimes an Italian explorer unable to find financial support back home could even persuade a Spanish king (probably competing with the Portuguese) to bankroll an expedition.

Evolution is full of stories like this treasure fleet tale. What species gets there first can make an enormous difference for a long time afterwards. And getting there first depends on whether populations are unified or fragmented. This is one aspect of what Stephen Jay Gould likes to refer to as "contingency" in evolution. Grand principles are perhaps there to be found (such as the six essentials for a Darwinian process), but they are not likely to substitute for the hard-to-learn details about the happenstance of history. You need both to make sense of things.

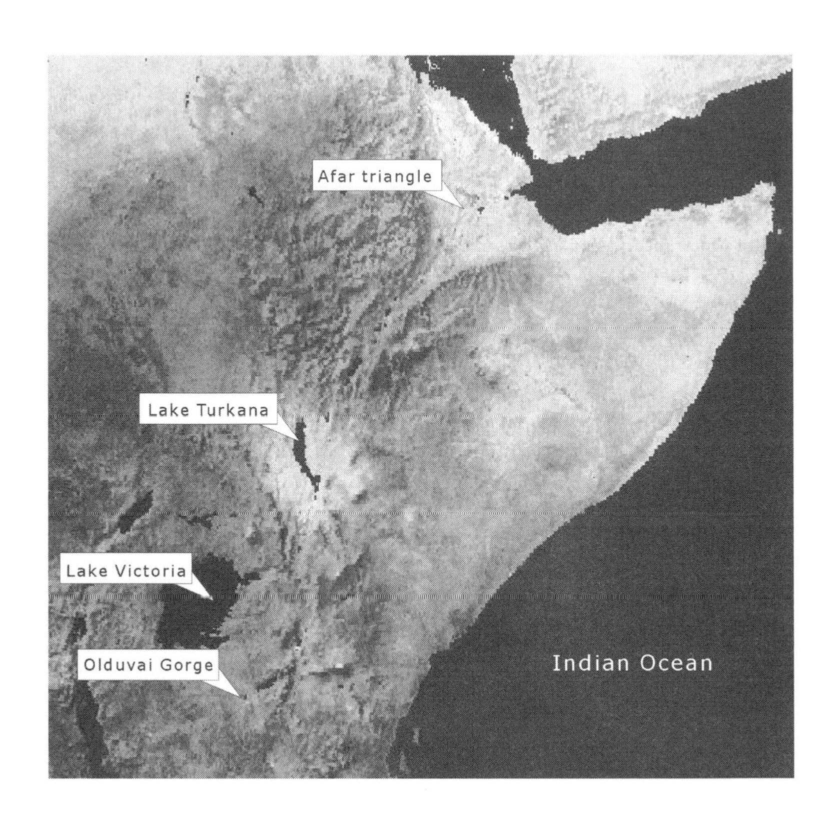
Afar triangle
Lake Turkana
Lake Victoria
Olduvai Gorge
Indian Ocean

To: Human Evolution E-Seminar
From: William H. Calvin
Location: 1.28680°S 36.81486°E 1,680m ASL
Nairobi, Kenya

Subject: **Creating new species from old ones**

Let me also recommend Ian Tattersall's *Becoming Human,* which I've been reading on the plane and in the more comfortable Nairobi watering-hole settings given by the GPS coordinates (with five decimal places of accuracy in latitude and longitude, you can almost locate my exact table). As Phillip Tobias pointed out to me in Johannesburg, Tattersall likes to subdivide things more than most, being on one end of the splitter-to-lumper spectrum.

I too think it probable that there have been a lot more hominid species than are usually recognized, simply because speciation speeds up evolution by preventing backsliding, just like that rock sill when the tide goes out. Because human evolution has been unusually rapid, it pays to look for fast tracks amidst the more general possible progressions, and additional speciations might be part of the fastest track.

FOR ALL OUR TALK about evolution, we find it surprisingly hard to define a species. Alfred Russel Wallace, in answering some of Darwin's critics who were confused about what a species is, emphasized that variability wasn't unlimited, that donkeys and horses each vary only within a certain range, that there isn't a continuum between them. Donkeys and horses may try to interbreed, but conception may not occur, or spontaneous abortions may prevent most live births. Those offspring that grow up may, like mules, be sterile, so that they don't contribute to maintaining some new middle ground between the two species.

But paleontologists cannot measure reproductive continuity like the zoologists can. They're stuck with, "Well, it looks different than this other one." Much confusion arises from these fundamentally different connotations of the word species. The fossil record only sees the anatomical differences, and so that's what we mostly talk about, when naming things. We constantly argue at cross purposes.

Of course, we know full well that some species have enormous regional variety but are still able to interbreed. Paleontologists digging up a pet cemetery would define dozens of domestic dog species that way, a great overestimate given how well they still interbreed, yielding short-haired mongrels. Then there are populations that look identical but cannot interbreed with much success (they have some behaviors that keep them apart, or perhaps chromosomal peculiarities). They are two different species but the paleontologists would count only one.

Many people working on chromosomes or DNA sequencing tend to talk about speciation in terms of major rearrangements, say the consolidation of the 24 pairs of chimpanzee chromosomes into our 23 pairs that occurred somewhere along the line. Yes, a prehuman male with 24 chromosome pairs might not be very effective at impregnating a more modern female with 23 pairs – but, like the focus on "mutations," this view of speciation vastly oversimplifies a lot of more relevant evolutionary and population biology.

It is often said that "attractiveness" would be important, that the behaviorally-modern people invading Europe might not have found Neandertals suitable mates. An exchange that I overheard at a physical anthropology meeting bears on this. Genetic model maker: "Even if only two percent of modern males attempted to mate with Neandertal women. . . . " Famous paleoanthropologist, interrupting: "Two percent?

More men than that will try to mate with a sheep! And they're not even closely related. . . ."

More effective as a behavioral isolating mechanism is a shift in the breeding season of a month or two (a behavioral change, perhaps with no anatomical correlate) between two regional populations of a species. That will keep two populations from effectively interbreeding when they meet again. The tassel-eared squirrels on the North Rim of the Grand Canyon, for example, breed in June, three months later than their brethren on the somewhat lower South Rim, where the snow melts in March. Should an individual from the north meet one of the opposite sex from the south on the hiker's footbridge across the Colorado River, they might not be interested in one another at the same time. Penned up together for a year or two, they might well hybridize, but the shift is pretty effective in maintaining their other differences (the North Rim squirrels have a skunk-like stripe) from dilution.

What speciation achieves, then, is the shifting of... the genetic and morphological centers of gravity of the parent and daughter species . . . Each species is now free to accumulate more variation and hence more potential species difference. Their descendant lineages will thus ratchet away from each other in a variety of directions. . . .
– IAN TATTERSALL, *Becoming Human*, 1998

Accumulating some physical differences is much easier in a small population. The local environment can really "select for" those variants that fit its peculiarities. Similarly, sexual selection's peculiarities like peacock tails can also get going most easily in small situations. But when the climate improves, and some immigrants arrive from a larger population elsewhere, this dilutes whatever adaptations that might have been achieved locally (one definition of "progress"). "Genes for cooperation" might have increased to involve half of the small population but then the percentage backslides with the dilution. The locals become more average.

SO YOU HAVE TO DISTINGUISH between the evolution of physical differences in a regional population ("adaptation") and the reproductively isolating mechanisms which occasionally preserve those physical differences ("speciation") from the usual back-and-forth mixing. Adaptation is the crank, speciation is the ratchet which prevents adaptation from backing up. Adaptation usually doesn't produce speciation, but sometimes they coincide. It's much like trying to cross a two-way street, waiting for gaps in the traffic. A gap in one lane doesn't help much, so you await simultaneous gaps before crossing.

It is when adaptation and speciation coincide that life undergoes sustainable change (though most new species, like most other small populations, promptly die out the next time the climate fluctuates). While isolation can allow regional physical differences to accumulate, it is reproductive isolation ("speciation") that protects these features from dilution, much as a ratchet prevents backsliding.

Even if they had viable offspring, there is a chance that such hybrids were sterile (in mammals, male hybrids such as mules are almost always sterile). But the question of successful interbreeding mostly hinges on the spontaneous abortion rate – nearly a quarter to a third of human conceptions are presently lost, mostly before a woman suspects she is pregnant (and this doesn't count the similar numbers of induced abortions). Drinking the wrong tap water, or too much caffeine, more than doubles the hidden abortions (fivefold in some studies).

Human pregnancy failure rates are surprisingly high, compared to domestic animals, enough so that it might have the same effect as a speciation, at least for awhile. If interbreeding attempts between the local populations – say, Neandertals – and the latest African emigrants had an even higher failure rate, few offspring would have been born. This raises the possibility that our ancestors might have easily "speciated" in the sense that immigrants to a locally adapted subpopulation – the ones that usually dilute out whatever "progress" has been locally achieved - might not have been able to interbreed effectively because of high local spontaneous abortion rates, just from drinking the wrong water or eating too much of the local plant toxins. This quasi-speciation would have the effect of saving valuable local adaptations from the dilution effects of mating with immigrants.

ONCE TWO POPULATIONS become reproductively isolated ("the species has split into two"), then they tend to compete with one another. Yes, they might ignore one another like two ships passing in the night, just a wave in passing, or they might even cooperate in some matters as several hyrax species do, but if they utilize much the same resources (food, nesting places) and have similar predators and parasites, then one of the species is likely to fare somewhat better than the other.

That's all that "competition" really means - though, of course, competition between populations can also include the bloodier forms (as closely related chimpanzees proved at Gombe, even without speciation separating the neighboring groups).

Note that a species (an inbreeding *population*, now) competes with other species within an ecosystem, rather like an individual competes with other individuals within a species. The species gets started, thrives, maybe splits off new versions, and likely dies out at some point. Natural selection may operate on how individuals thrive and die, but it can also shape which species thrive.

While selection may mostly operate on individual genes within the genome, there is a sense in which the performance of the entire committee of genes is also under selection, as some committees do better than others. Indeed, it is this collection which serves as the source of new variation, what continues to look for better fits with the environment. This committee aspect is most obvious at the cell level: the genes within either work together or die together as the cell membrane ruptures.

Darwin's inheritance principle operates at the species level as well, although it doesn't have quite the same "hang together or hang separately" aspect that the cell level has. Still, it is the species which encompasses the full range of ancestral "good tricks" that can be used to survive and thrive in a new climate regime. Having the right alleles (different versions of a given gene, as when the A1 allele of the D2 dopamine receptor produces 30 percent fewer receptors than the more common A2 allele) available somewhere in the population can make one species do better than another when the climate perturbs things.

THERE IS A STATUE OF LOUIS LEAKEY outside the research building at the National Museum of Kenya, which takes a little work to locate, as it is located back behind the left side of the museum. Louis is shown seated, contemplating a handaxe. I like to think that he looks a little puzzled by this enigmatic tool.

There are a lot of anatomical differences between us and all those ancestral bones that I saw today at the museum. As Tattersall notes, each little change probably had to be insulated against backsliding, using yet another instance of reproductive isolation. In other words, a new (physiological) species. That ratcheting makes it likely that there have been a lot more physiological species (say, dozens) than the half-dozen ancestral anatomical species we usually argue over.

I can say this so simply, but there has been a century of argument about species and the pace of evolution where nothing was simple. The debate isn't finished yet.

Evolution is an enchanted loom of shuttling DNA codes, whose evanescent patterns, as they dance their partners through geological deep time, weave a massive database of ancestral wisdom, a digitally coded description of ancestral worlds and what it took to survive in them.

– Richard Dawkins, *Climbing Mount Improbable*, 1996

To:	Human Evolution E-Seminar	
From:	William H. Calvin	
Location:	1.57816°S 36.44417°E	1,000m ASL
	Olorgesailie, Kenya	

Subject: **The easiest tool of all**

We were driving along a nice, flat stretch of road, and I looked around to see that both sides of the road were flat for a long way back. Aha, I said, drawing it to the attention of the expatriate cousin. We are, I announced, in the midst of another former lake bed.

Lakes shrink and expand as droughts come and go, so shorelines migrate. There's another kind of change, too: most shallow lakes silt up eventually, if something doesn't pull the plug before then. That's why it's so flat here. There's a wonderful example to the west of Wilson Airport in Nairobi, a light-green-surfaced "pond" with dark-green woods surrounding it - except that the "pond" now has a road running across it. And it's probably not a floating pontoon bridge, either.

We're on our way to visit Olorgesailie, which is right on the edge of a basin that held a lake until it lost part of its rim to earth settling. About a million years ago, it contained Lake Olorgesailie (a quite large freshwater lake, over a hundred km^2, now vanished), much like the still-wet ones in the Rift Valley north of Nairobi that we'll visit starting tomorrow. When you fly over this area southwest of Nairobi, you see a series of dry basins, each of which was once likely a big lake. It almost looks like hillside terraces forming a series of modern rice paddies, cascading down to Lake Magadi.

I CAN'T VIEW OLORGESAILIE without thinking of all the time that famous anthropologists have spent here. Back during World

War Two, Louis and Mary Leakey were largely confined to Nairobi by war duties and the petrol shortage, but they organized weekend digging expeditions down to Olorgesailie, about 90 km south of Nairobi on the Lake Magadi road in the midst of an acacia savanna. They'd invite along their war-weary friends and put them to work, moving rocks. Later in the war, some Italian prisoners-of-war helped out (one later became the warden of the national park created at the site in 1948). Then in the early 1960s, Glynn and Barbara Isaac spent a lot of time digging here, leading to their classic monograph, *Olorgesailie*.

Glynn Isaac had a dramatic way of lecturing about early toolmaking. On the way to the lecture hall, he would pick up two fist-sized rocks from whatever offered locally. He would place them atop the lectern and let the audience wonder what they were for, while he showed slides that introduced tools and their variety. Then he would claim that the simplest sort of toolmaking would yield the most useful implement of all, the equivalent of the single-edged razor blade, handy for incising hides and amputating joints.

He would put on his glasses, take a rock in each hand, turn his back on the audience, and begin hitting one rock against the other. Chips would start flying toward the rear of the stage. Glynn would keep this up until there were several dozen fragments scattered about. There was nothing delicate about his technique. It was just sheer brute force, with only enough sense of aim to keep from hitting the hand that held the target rock.

Then Glynn would stop. Silence, after the fury. He would lay down the two remaining half-rocks and sort through the fragments littering the stage floor. Ah-ha, he would say, here's a sharp one. Holding it up, testing its edge, he would proclaim it just the thing for butchering a zebra. With it, you could take off a leg at the knee, then run away with dinner. He would look at the remaining half-rocks and proclaim one of them good enough for defleshing – good solid rounded surface to grip, but still enough of an edge to hack away and separate skin from flesh.

What more could you want? And all from a minute's unskilled work. Well, you needed enough skill to prevent hitting your own hand. But you could get around that, for a true beginner's technique, by simply throwing a rock against a hard surface. It was so simple that even an ape could probably do it.

INDEED, ONE CAN. I once spent two lovely weeks in 1990 on the Portuguese coast discussing tools and language with several dozen diverse scientists. Sue Savage-Rumbaugh was there, showing videos of what Kanzi the bonobo could do with words, and Nick Toth gave us a nice toolmaking demonstration on the beach. One of the aftermaths of this Portugal meeting was that both Nick and I visited Kanzi in Atlanta, not long afterwards. When Sue told me that Nick was going to demonstrate toolmaking to Kanzi, I told her about Glynn Isaac's demo and joked that Kanzi would discover the beginner's technique of brute force fracturing and sorting through the fragments.

And that is indeed what subsequently happened. Though Kanzi tried to imitate Nick's more careful technique, he eventually threw the rock against the concrete floor and got a sharp flake that way, using it to saw through the rope that held shut the box with the food reward. Sue then moved the scene outdoors where there was soft ground, to discourage this simple technique and try to get Kanzi to improve his rock-hammering technique. He is now rather good at it; though still lacking a good hominid sense of "the right shape," he can tell from the sound whether he has produced a flake which is sharp enough. Thus at least one great ape is capable of the entry-level, what hominids were doing starting 2.6 million years ago.

So the earliest tools weren't even designed. They were closer to found objects, though generated by a bit of banging around. And that simple procedure yielded an enormous improvement in what could be eaten.

THIS PROBABLY WASN'T THE BEGINNING of meat-eating, of course. Judging from baboons and chimps, there is a lot of opportunistic snatching of various small mammals (usually the defenseless young) who are hiding in the tall grasses. And chimps are rather good at group maneuvers which trap small monkeys. Chimps clearly love fresh meat and fat (the monkey's brain is the most-sought-after prize).

Though chimps will usually pass up a dead monkey left on a path for them to find, scavenging on the savanna has some not-so-obvious rewards that our ancestors might have utilized. The bigger predators, such as the big cats, will eat their fill of soft parts, without chewing too much limb muscle (they too go after the fat first, both abdominal and periorbital). The jackals and hyenas may take much of what's left. You'd think that there wasn't much left for a hominid unable to drive off such fierce competitors. But there is a lot of bone marrow, well hidden, if you just wait for the others to leave. Hyenas have jaws strong enough to crush many bones, so they are your main competition for the marrow. Throwing stones to drive them away is one possibility.

Breaking open the long bones can be done by pounding hard enough to create a torsional fracture. This opens up a length of bone shaft. The marrow comes out looking much like a long pink sausage. So this scavenging technique can make good use of the leftovers. Just bring along a rock that is heavy enough – or, even easier, cut off the leg at a joint with one of Glynn's sharp fragments and swing the leg like a club against a tree or a rock outcrop. Amputating a leg also allows you to quickly carry the remaining meat and marrow to a safer place, away from the sharp-toothed competition. The clublike fracturing technique can even be practiced high up in a tree, where heavy pounding is more difficult. There are some archaeological sites that have an excessive number of long bones, complete with appropriate fractures.

But once such basic discoveries are made, scavenging isn't much of a growth industry. You are still dependent on big cats and the hyenas to make your kills. So in the familiar logic of food pyramids, your population numbers will always be limited to only a fraction of theirs. Just as top predators are rare, so are top scavengers. Is there a way of expanding prehuman meat-acquisition abilities, beyond that seen in chimps and beyond this "myeloscavenging" for marrow?

Yes. Surely several. I'll describe one after I do a little scene setting. But note that tool *use* per se need not leave an archaeological record.

OLORGESAILIE SHOWS the Acheulian toolkit that appeared about 1.8 million years ago. It didn't change very much for a million years thereafter. In toolmaking, there didn't seem to be any steady progress, contrary to our "Man the Toolmaker" expectations of what drove things. The initial toolkit seen about 2.6 million years ago in Ethiopia had no more than a half-dozen tool types, and the Acheulian toolkit has no more than a dozen. But in contrast to what came earlier, the Acheulian toolkit has at least one tool that looks "designed" (though for what isn't clear).

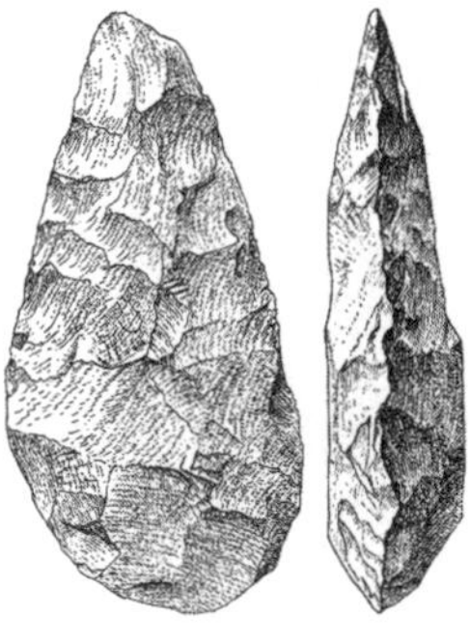

This biface is a rock that has been symmetrically shaped into a horizontal plane (some are so thin as to look like a frisbee). But it isn't round. It looks somewhat like an arrowhead, bilaterally symmetric around a point. Except the point isn't sharp, and the back end isn't hafted for a handle or arrow shaft. Indeed the back end is often (though not always) sharp, just like the front and sides, sharpened by carefully chipping away at both faces. When the back edge isn't sharp, the name

"biface" is usually applied and that's perhaps the best name for all of them, given the current misleading nomenclature.

As I mentioned back in Paris, the classic sharpened-all-around version tends to be called a handaxe. The classic "Acheulian handaxe" is what you see at Olorgesailie by the thousands. It's like being on a cobble beach with lots of fist-sized round stones – except that most of the stones here were obviously reshaped by hominids who were obsessed with something. Old lake beds in the Sahara, uncovered by sand dunes that shift sideways, can be similarly littered with handaxes. There's a dig at Olorgesailie which Louis Leakey called the "factory site" that testifies to the huge quantities left behind over the years. But why?

Remains of zebras, monkeys, antelopes, giraffes, hippopotamuses, rhinoceroses, and elephants are all in the ground at Olorgesailie, which may suggest that past and present savanna habitats are very much the same. However, about half of the fossil mammals at Olorgesailie no longer roam the savanna; they are extinct. Their council included a frighteningly huge species of monkey, a large form of zebra, at least two inordinately bulky pigs, a disturbingly massive species of elephant, a few kinds of grass-eating antelope, and a heavier-than-usual hippopotamus.

Everything we know about them indicates they were products of the Cenozoic decline, dedicated to eating grass. Their cheek teeth were large and specialized for chewing abrasive blades of grass and herbs. Their bodies were larger than their ancestors' and more massive than equivalent forms on the present savanna. These anatomical facts mean that they took on the quintessential approach to the Pleistocene plains: Eat grass, and eat it abundantly. I tend to think of them as huge lawnmowers.

– RICK POTTS, *Humanity's Descent*, 1996

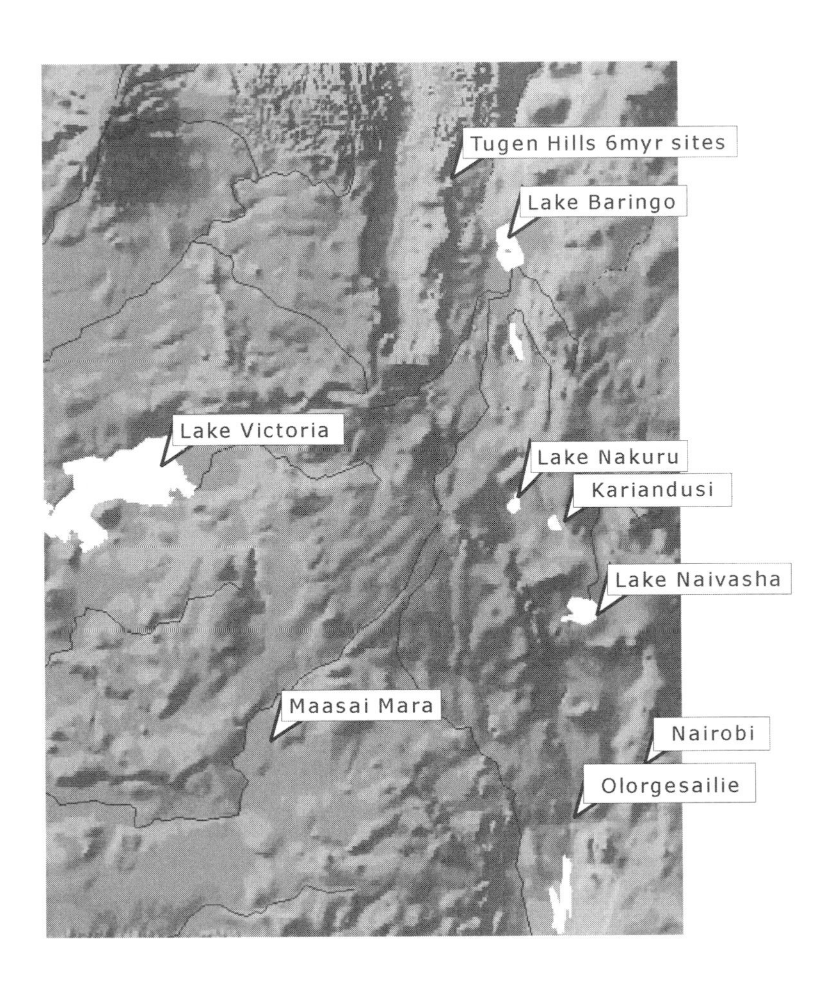
Tugen Hills 6myr sites
Lake Baringo
Lake Victoria
Lake Nakuru
Kariandusi
Lake Naivasha
Maasai Mara
Nairobi
Olorgesailie

To:	Human Evolution E-Seminar
From:	William H. Calvin
Location:	0.45187°S 36.28296°E 1,890m ASL Kariandusi, Rift Valley

Subject: **A layer cake of handaxes**

The Rift Valley is beautiful and amazing, well beyond my powers of description. Driving north from Nairobi on the old road is spectacular. We are now, thanks to some advise from Meave Leakey, well above the present shoreline of Lake Elmenteita near a quarry. The lake level used to reach up past here, in the good old days, hundreds of meters higher than the present levels of Lake Nakuru and Lake Elmenteita.

And this ancient shoreline is what made Kariandusi into another version of what we saw at Olorgesailie yesterday: here, too, there is a Leakey-excavated sea of handaxes. Except that here, you can easily see one layer atop another. And then another, like a layer cake. My cousin, primed by seeing Olorgesailie and now knowing what to look for, is gratifyingly amazed. Kariandusi is about the same age as Olorgesailie, with estimates ranging from 700,000 to a million years old, right in the middle of *Homo erectus* time. Together, Olorgesailie and Kariandusi give you a wonderful lesson in why *erectus* toolmaking is so enigmatic. And why climate changes moving lakeshores might be so central to understanding *Homo erectus*.

As at Olorgesailie, there is a fine one-room museum at Kariandusi, another branch of the National Museum of Kenya which runs this site. The museum here has the same hominid skull casts seen in Nairobi (and in museums around the world). It also has a nice short version of how the Rift Valley formed in geological time, the floor dropping down between two cracks (the two escarpments we see mark, naturally, the

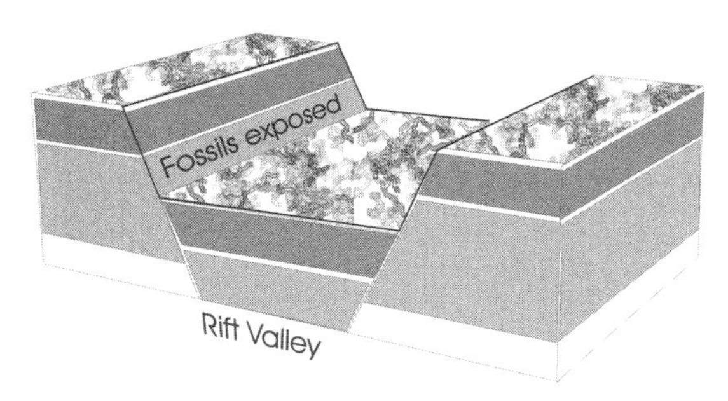

two big north-south fault lines) as the rift widened. The walls of the valley have a lot of fossils exposed in some places.

The museum appears to train the guides well. If some of what they say about handaxes is implausible, it's only because they are faithfully repeating what archaeologists have said over the decades (sometimes, one suspects, with tongue in cheek).

THE KARIANDUSI MUSEUM has a poster full of the proposed uses of the handaxe. It shows you the extremes to which archaeologists have been pushed, trying to find some definitive use for the strangely-shaped handaxe which is the defining hallmark of *Homo erectus* culture. They all have uses for it - and they're probably mostly correct, too - but none adequately explain why its shape is what it is, or why that shape escapes the usual cultural drift for close to 1.8 million years.

Many modern archaeologists are suspicious that they might be dealing with a blind-men-and-the-elephant situation when it comes to the handaxe, that they possess only part of the truth and are making erroneous extrapolations. They just don't know what to replace or supplement them with. Because of the history of inadequate explanations, most have become totally skeptical of any new explanation that is offered for the handaxe, even by fellow archaeologists. Been there, done that.

The handaxe poster starts with "For butchering." Yes, one edge of a handaxe could be used for clearing meat from the bones. But why is the back edge sharpened too, the one the hand has to wrap around? I've never heard an answer to that one. And why use an elaborate tool when a simpler one usually suffices? Why use such a symmetric, hard-to-make tool for a purpose that one of Glynn's half-rocks or random flakes would suffice for? It's overkill, as the phase goes. The handaxe sure wasn't designed for just defleshing.

"For digging up edible roots" reads the next panel. That's even more difficult when the implement is sharpened all around the perimeter, because you tend to pound on roots. You'd break many handaxes that way (not to mention cutting the palm of your hand). And, in any event, overkill again. There are some sturdy bifaces called "picks" that might stand up to cutting roots loose, but here I am concerned with the classic, enigmatic handaxe shape, not what broken or tumbled versions of it might secondarily prove useful for.

"For scraping animal hides." (Clothing, so early? Well, just assume they mean skinning the beast to get at the limb meat.) I'd sure prefer something with a comfortable grip, myself. And one of Glynn's flakes or half-rocks would work pretty well for skinning. It doesn't even have to be particularly sharp for skinning, so long as you maintain tension by lifting the hide loose as you separate skin from fascia. When I learned my surgery, one of the big surprises the first week was how handy dull tools were, that you often wanted to use the dull side of the scissors to spread, or the blunt handle of the scalpel to scrape, rather than the sharpened surfaces. Only beginners think that the scalpel is the quintessential surgeon's instrument. Once you get through the skin somehow, surgery is mostly blunt dissection, spreading and separating along natural planes, such as between fascia and skin, where a sharp edge is a nuisance most of the time.

In one excavation at Olorgesailie, our team uncovered a skeleton of the ancient form of elephant known as *Elephas recki*. It was surrounded by hundreds of sharp lava flakes that, nearly 1 million years ago, had been used to butcher it. The flakes were boldly struck from handaxes, yet not a single intact handaxe was left behind. The handaxes were blade dispensers.
– RICK POTTS, 1996

"As a core from which tools were struck." That's a good possibility, at least for the larger handaxes that lack fine retouching and show cavities where a flake used to reside. But it doesn't explain the bilateral attempt at symmetry, or the tendency toward a point.

"As an anvil." Now there's a trivial one. Any flattish rock that was carried around would do, of course. No design needed. But if you were carrying around a designed rock that was flattish and sturdy enough - sure, why not use it as an anvil, should you need to soften up a root or crack a nut?

"Thrown discus-style as a weapon." Now we're talking. Throwing has great possibilities in human evolution. But the museum's artist got it all wrong, as the cartoon shows an overarm throw at a lone antelope, the handaxe's pseudo-point conveniently facing toward the animal, miraculously stabilized in that unlikely orientation without the aid of a spear shaft. But at least "discus-style" was included in the caption, what I think is the key to the puzzle.

SO YOU CAN SEE WHY it's been called the "Swiss Army Knife of the Paleolithic." If no single use can justify its design, maybe it is a composite tool! Or so the story goes (that phrase "jack of all trades and master of none" comes to mind). Yet, ironically, the name "handaxe" is a misnomer because using it as a handheld axe is the least likely of its proposed uses. In its classic form, it is sharpened all around its perimeter. It is thus likely to bite any hand that held it, if striking something with any force. So put scare quotes around "handaxe" from here on, whenever I forget them.

Furthermore, almost none of the proposed uses account for its distinctive shape. Certainly not simple butchering. It could easily have been used as a core, from which smaller flakes were made, and there is some evidence for this use. Unfortunately, many handaxes have so much fine finishing – tiny chips removed here and there – that it is clear that such handaxes were something more than merely a core for thumb-sized flakes. Yes, hacking, scraping, and cutting are possible uses (and there is even some edge-wear supporting woodworking) – but the handaxe wouldn't have done those jobs better than Glynn's broken rocks with a fortuitous edge. And they (and the Oldowan tools) are sure a lot easier to make, and they all continued to be made (the handaxe didn't replace them).

And why is the handaxe flattened and bilaterally symmetric, with a suggestion of a point? (It is, however, seldom sharpened to a penetrating point, the way you see in the spear tips and arrowheads in the last ice age.) None of the Swiss Army Knife explanations ever get around to addressing such handaxe issues.

It would be difficult to over-emphasize just how strange the handaxe is when compared to the products of modern culture.
– THOMAS WYNN, 1995

Even worse, this classic shape persists for nearly 1.8 million years, even into Africa's Middle Stone Age about 100,000 years ago. It is seen in Africa, Europe, and now even in China. It stays about the same shape, regardless of what local rock is used. That really suggests we're missing something important. To conserve the classic form through the vicissitudes of cultural drift and local fashion, there must be something that keeps the shape drifting back toward some functional optimum.

But for what function? It's surely not the Swiss Army Knife functional collection, as combination devices are surely subject to drifting styles, with features being omitted when they are no longer supported by the culture.

I suspect that I wouldn't have told you all this if I didn't myself have a candidate for what the prime use was, the one that none of the Swiss Army Knife uses adequately addresses. It builds on some experiments that Eileen O'Brien did back in 1979, the most unsettling undergraduate thesis I have ever read. And since it meshes nicely with the waterhole aspect of the savanna story that I'm trying to get across, perhaps I should tell the updated version of my handaxe story.

WHEN I FIRST HATCHED my handaxe scheme back in the 1980s, someone immediately attached the name "Killer Frisbee" to it. Just try throwing a handaxe, even with the initially horizontal plane of rotation favored by most modern Frisbee throwers, and it will soon turn on edge and come down vertically. Happens even to experts, not just me (I repeated O'Brien's experiments a decade later, also employing a well-trained discus thrower). Doesn't matter how you throw it - overhand, sidearm, underarm - the same thing happens. "Handaxes" are vertical-plane spinners.

Unlike round Frisbees, a handaxe will not lead you on a merry chase after landing, rolling along forever. The handaxe's "point" will shortly bury itself in the ground and bring the implement to a shuddering halt. Is this good for something? Some context must be missing, and it's not likely to be a prehistoric game of paleofrisbee.

The context, I suggested back at that Portugal meeting, is when throwing such an object into a tightly-packed herd visiting a waterhole, especially during a drought when their choices in watering sites are very restricted and they try the safety-in-numbers strategy. Then the puzzle parts seem to fall together.

I don't think for a moment that the handaxe was a general-purpose hunting projectile (the archaeologists are rightly skeptical of that use). I think that you can understand the handaxe only in the special-purpose context of closely-packed herds at the waterhole.

LET ME START with a few words about shoreline predation, surely a topic of general importance for hominid evolution, not merely the handaxe enigma. The big cats, as well as various other carnivores, have all discovered that lakes, rivers, and waterholes are a good place to hang out. That's because the grazing animals eventually have to visit a water source. Wait for the animals to come to you. Saves chasing them, and all that effort. Hominids would have realized the same thing. (Recall what the bank robber said when asked about why he robbed banks: "Because that's where the money is.")

Of course, after a few successes, the herd gets a little wary and picks another watering site. But sometimes there isn't another choice. Some waterholes are obligatory, as all the others within walking distance have become mud holes, suitable only for elephant beauty baths. And so each evening, the herd, packed tightly together for mutual protection, comes down to the waterhole. The tails of the waiting predators begin to twitch in anticipation. Sometimes there were tailless predators instead, of the hominid variety.

Being on the outer rim of the herd is not the best place to be. Packing tightly does reduce the percentage of the herd that is exposed to predation. The bigger the herd, the bigger the perimeter - but the perimeter grows more slowly than the area, and so, in the larger herds, only a few percent are actually exposed to an attack. With only a dozen animals, half of the herd is exposed.

An upright hominid might have been able to get closer to the herd than the familiar four-legged carnivores. But, as I earlier mentioned, the two-legged advantage doesn't last long. The herd will increase its approach distance - the distance from which they start moving away from you, if not running away. And besides, what's the hominid going to do, anyway - get close enough to club a hapless animal?

So there's the scene: tightly-packed herd visiting waterhole, two-legged wannabe predator standing at some distance, without even a tail to twitch. What happened next?

THE ENTRY-LEVEL SOLUTION, I said back at the Portugal meeting, was to throw a tree branch into the midst of the herd (chimpanzees like to wave and fling branches, so maybe our ancestors did too).

The herd will panic, of course, wheeling around to flee. The branch lands in their midst (towards the rear is best), perhaps hitting an animal, perhaps not. But soon an animal trips over it or gets entangled, knocked off its feet by adjacent stampeding animals. It tries to get back on its feet but another knocks it down in the confusion. Pretty soon it gets halfway back on its feet, but it is too late. Two-legged predators arrive and grab it. Maybe they club it, maybe four just cooperatively hold a leg each in the chimpanzee manner while a fifth chews on its throat. One way or another, they make a meal of it.

The bigger and more tightly packed the herd, the better the technique works. The usual surface-to-volume ratio advantage of herds and schooling is turned on its head by this simple invention, of lobbing a branch over the top and into their midst. The bigger the herd, the easier it is to disable one in the middle.

Pretty soon, herds grow wary – but they form bigger and more tightly packed groups in response, making the technique work even better. The local trees are denuded of convenient branches. While waving and flinging branches are clearly favorite scare tactics of chimpanzees (they have not been seen to use accurate throwing for predation per se), they will resort to clods of dirt and rocks when branches are in short supply. Rocks, the hominids might have noticed, will go farther than anything else. The thrower can launch while still outside the approach distance. But what does the rock buy them? Few animals will trip over it when fleeing. With a beginner's throw, knockouts are unlikely.

Ah, but the herd is, remember, tightly packed in this scenario. Even with the beginner's side-of-the-barn throw, a lob can't miss hitting *some* animal somewhere. If the rock is big enough and the animal small enough, even a rock landing on the flank will bowl over the animal. Were it not for the rest of the herd running past at close quarters, the downed animal could likely get up and run away in time. Few would be injured enough to stay put.

But the herd is running past, and attempts to get back up in time may fail. Another hominid meal is procured.

Rocks are more reusable than branches, which get trampled into mere sticks. Are some rocks better than others for this tightly-packed-herd use?

SUPPOSE YOU HAD AN OUTCROP of sedimentary rocks that was layered, so some rocks were pre-flattened in the manner of slate. Throw one of those and you'd get some of the frisbeelike aerodynamics. Goes farther. Lands on edge, too. Sometimes on an animal's back.

That has an interesting neurological consequence. All mammals have flexion reflexes, used to withdraw the body part from further harm. If you step on a thorn, you quickly withdraw your foot and shift your weight to your other foot, all before your brain even takes notice of the thorn. Most archaeologists know this, but what they might not know is that withdrawal reflexes also work for damage to other parts of the body - such as the back or flank. If a grazing animal felt something sharp on its back (as from overhanging thorn trees, with acacias common hereabouts), it would tend to sit down quickly. The big cats may have discovered this reflex tendency to crouch, given how they leap onto an animal's back and dig their claws in.

So flat stones with sharp edges might get a reputation for being more effective. If you are also knocking off sharp edges from such a stone to butcher the animal, it might have some sharp edges for next time. Two uses for the price of one.

The preferred projectile is now basically flat with sharp edges, somewhat symmetric for less tumbling en route. But it doesn't necessarily have a point. Why add a suggestion of a point, without making it really pointed for penetration?

Some would have been semi-pointed, of course, perhaps as a consequence of all the flaking used for butchering. Or perhaps just from breaking off a segment (some would be trampled by the herd, after all, and then recovered). And the hominid hunters might have noticed that a bit of a point made the stone more effective in bringing an animal down. (You don't have to understand the flexion reflex to make an empirical observation about effectiveness.) Some of the stones would just bounce off the animal's back without toppling it. But those flat rocks with a bit of a point, given that they are spinning, will soon embed their pseudopoint into a pushed-up roll of hide.

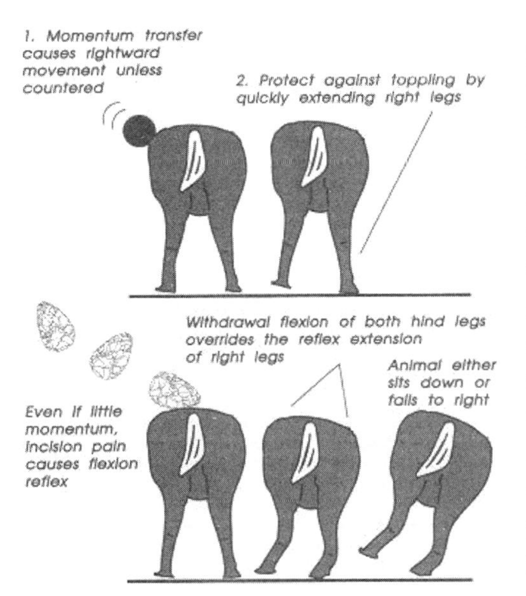

There's no need to penetrate the skin. Why does it work better at toppling the animal? Because it stops the spinning stone, before it can bounce free. Most throws will not strike the flank solidly, transferring all their momentum to the hapless animal. Most will bounce off the animal's back, slowed only slightly. But with the pseudo-point catching the roll of hide, all of the stone's momentum is transferred to the animal, rather than the usual fraction.

Because of that, even palm-sized handaxe-shaped stones may be effective at toppling the animal. The momentum transfer gets the animal moving slowly sideways, and the pain's flexion reflex (also enhanced by the abrupt jerk when the point catches a roll of skin) interferes with the farside leg extension that usually prevents toppling. Over it goes, ever so slowly, knocked off its feet.

Thus the classic features of the handaxe - somewhat flattened, mostly edged, mostly symmetric, somewhat pointed - ought to be one of the nice solutions to this hominid version of the waterhole predation game. They'd surely retrieve and reuse favorite projectiles, of course. However, given how deep the mud is around those waterholes, they would probably lose a lot of them, too. Over the years, quite a few handaxes would accumulate in the shoreline mud. That's surely a better explanation for accumulations like those at Olorgesailie than that the sites were some sort of factory, doing an early form of mass production of handaxes.

It does suggest that the flattened teardrop is a pragmatic shape, not necessarily some innate symmetry in the mind of the toolmaker. Even the more-or-less spherical "hammer-stones" were shown by the archaeologist Nick Toth to be a consequence of repeated use, not design. Any stone tends to become round, once you hammer with it long enough.

The Killer Frisbee use might also help explain why a few handaxes are found standing on edge in old watercourses and lake edges. I doubt the handaxes were planted that way, just to serve as primitive sundials or to confuse latter-day archaeologists. Thrown handaxes always land on edge, digging themselves a nice groove and often planting themselves in a vertical orientation, though most surely topple over during subsequent erosion of the landscape, as when they are tumbled by stream flow.

THE LAYERS HERE at Kariandusi make my point even more dramatically than the sea at Olorgesailie. In each layer of old lakeshore silt and mud, there are hundreds of hand-sized stones showing signs of prehuman modification, most of them classic handaxes. The next layer a few centimeters below (exposed a little farther downhill) and you see the same thing again. And again. It reminds me of the old jailhouse story of sending a prisoner a cake with escape tools baked inside. Well, each layer of this "cake" is loaded with an unbelievable number of teardrop-shaped tools.

If the waterhole predation theories of handaxe use are even partially correct, there ought to be a series of concentric rings of handaxes, layer by layer down towards present-day Lake Elmenteita, shrunken from its former self. Well, perhaps not whole rings, as only certain sites on the lake's perimeter would be favored by both animal trails and hominid hunting habits, but in any event there should be a lot of sites, given a million years worth of *Homo erectus* families to feed.

Some Kariandusi handaxes appear somewhat smoothed, as if tumbled in flowing water - and their edges wouldn't bite the hand that held them, making it possible to use them as a handheld axe of sorts. But adjacent ones may have sharp edges, as if never moved downstream after being lost. Hmmm. Remember that floods happen, and climates eventually change.

WHEN A RIVER FLOWS FASTER than usual, it can cut a deeper channel and so strip off layers of sediment from its banks and bottom. New river banks appear, showing off layers of smoothed cobbles from several earlier river beds. These are a common sight to anyone who hikes along a river today and looks at the river banks, and it would have been much the same a million years ago. But here, some of those rock layers wouldn't be just cobbles. Guess what might be left exposed after all this down-cutting?

The river banks and bottoms would contain overwhelming numbers of symmetric, pseudo-pointed rocks. Some such "handaxes," of course, would have been eroded out and carried downstream and tumbled smoother. Such tool-studded river banks would become known as Handaxe Heaven in whatever language the locals had attained. Would these early humans have called it "manna from heaven" or ascribed this bounty, perhaps, to their ancestors having provided for them?

It lends a whole new layer of interpretation to *objet trouvé*. (I can hear it now: "No, this tribe didn't actually *make* handaxes, they merely *recycled* the lost ones of an earlier culture. Recycling is a million years old.")

And it shows you a wonderful route to reinvention. Techniques are always getting lost, much as the Australian aborigines in Tasmania eventually lost some fire-making and tool-making techniques over the millennia, once rising sea level had isolated them from the mainland. Cultural loss is one of the hazards of the population downsizing and fragmentation that climate change can cause. Riverbank hunters, however, with no more than just the chimpanzeelike tendencies to lob rocks when a branch wasn't handy, could easily rediscover the aerodynamic properties of the handaxe and become superstitious about using the "right shape." When the provident riverbanks were depleted, they might start reshaping local rock into the classic shape. They'd already know, from long use, that the shape was useful. It's one reason why the classic shape might have survived a million years of climate change and all the island-like fragmentations of *Homo erectus* populations that repeatedly happened.

Mining "handaxes" likely also occurred from sites which were no longer riverbanks, sites like Olorgesailie and Kariandusi are today. You can consider it as either the beginnings of mining or of archaeology. "Making use of the past" is a modern museum motto but you can see how it might have gotten started.

There's another kind of reuse of handaxes, too. Consider the life history of the classic handaxe shape: if not lost first, most classic edged-all-around handaxes will eventually be broken, thanks to landing on something hard or herds trampling them. The handaxe might lose a third of its sharp perimeter. If thrown in the usual manner, it would tumble and not go as far, nor land on edge in the preferred manner. But the remnant's broken edge would be more amenable to a hand grip. Now you could use the remaining sharp edges for all of those Swiss Army knife tasks that the archaeologists proposed in their frustration: pounding roots, defleshing, and so forth. Ditto for the tumbled handaxes, smoothed to make them safe for the palm of your hand. So no-one even has to give up a favorite theory.

It's not often when you can have your cake and eat it too. Just remember that the handaxe has two lives, if not irretrievably lost in the lakeshore mud: Frisbee first, Swiss second.

To:	Human Evolution E-Seminar		
From:	William H. Calvin		
Location:	0.39804°S	36.08224°E	1,800m ASL
	Lake Nakuru		

Subject: **Where droughts cause a boom time**

I wouldn't have spent so much time on handaxes if it were not for how central predation strategies are, if you want to understand why *Homo erectus* could have thrived at the water's edge. Now let me turn to how climate change attracts herds down to the lakeshore.

Here at Lake Nakuru, as at the other Kenyan national parks, one is amazed at how many different species of mammal and bird can be seen in any ten-minute period. Besides the variety, sheer numbers are also striking. It isn't like riding the elevated train around the San Diego Zoo's Wild Animal Park.

What's so different about the real thing (besides one animal eating another, something frowned on at the best zoos) is the savanna setting itself. Acacia trees are everywhere in East Africa, each with a characteristic branching pattern rather like that of the apical dendrites of the pyramidal neurons of our neocortex. Maybe I ought to start a movement within neuroscience to rename them "acacia neurons."

Another difference in the Kenyan game parks is the relative lack of people inside. Outside, there are an amazing number of people but without the infrastructure to match. If you stop your car at a deserted pullout with a nice desert view, an entire family of Africans will appear within a few minutes, emerging from somewhere in the seemingly empty landscape. It certainly makes you cautious about saying that an empty landscape is truly uninhabited. At least in Kenya, the people are there, hidden somewhere in whatever shade offers. Their ability to make a living in such sparse environments is quite an accomplishment. But they live not far from the edge of famine, as any worsening of rainfall results in more people than food.

Which reminds me of what would happen if the Kenyan government became even more ineffectual than at present. Would-be anarchists and the minimal-government types really ought to try living for a few years in a country where the government is paralyzed.

A truly humane society would have no more people than can be fed in the worst years of climate fluctuation. But many tribal leaders, and some leaders of whole countries (say, Israel at various times, worried about the high Palestinian birthrates), tend to think that an increasing population gives them an edge over neighbors.

LAKES LIKE NAKURU show you a lot about climate fluctuations. The lake level itself varies a good deal, and thus the lake's size. The shoreline is shrinking right now, as it is in a number of large lakes in Kenya, and so you see the mud flat drying out. Grass starts growing there, supported by the water table, and many species arrive to graze on the new shoots, even if they have other water supplies. Fly over Chobe National Park in Botswana and you will see a series of bull's-eye targets: shining water in the middle, an intermediate ring of wet mud and grass, and then a dry ring on the outside. These ponds have animal trails connecting them, looking like a network with many nodes stretching across the landscape. Some of this drying up is seasonal (Amboseli, the ephemeral lake on Kenya's southeastern border, is famous for it) and some is longer-term climate change.

Furthermore, in dry times, the margin of a large lake ought to concentrate populations quite a bit. In good times they can spread out, but in a drought they all have to stick close to the rivers or a lake margin. So lake margins are refugia, places where a species can continue to find most of the elements of its niche – food, protection from most predators (recall those bush airstrip landscape esthetics favoring open views), tolerable levels of parasites, nesting sites and other elements needed for reproduction. And for a savanna-specialized animal like our hominid ancestors a million years ago, a drought would concentrate the game nicely. The drought even produces more savanna locally, thanks to that ex-mud-flat rim turning grassy.

What was a crisis for many species might not have been such a crisis for our ancestors, if they usually lived along waterways and lakes. A drought would force the game to come to them. After hominids spread away from waterways, a drought would have been a problem – but before that range expansion, a drought might have been a boom time.

A single dry year may kill off stock, reduce grazing land, and devastate crops, but improved rainfall the next year will mitigate the impact. However, a succession of arid years may have a cumulative effect on cattle and humans, to the point that an unusually severe drought can deliver a knockout blow to already weakened communities. Sahelian dry cycles can persist for up to fifteen years, as can periods of higher rainfall. The latter lulls everyone into a false sense of security. Cattle herds grow, fields are planted ever farther north into normally arid land, contributing to the disaster if a long dry period arrives without warning. If anything is "normal" in the frontier lands, it is the certainty that severe drought returns. The ancient Sahelian cattle herders planned their lives accordingly.

– BRIAN FAGAN, *Floods, Famines, and Emperors*, 1999

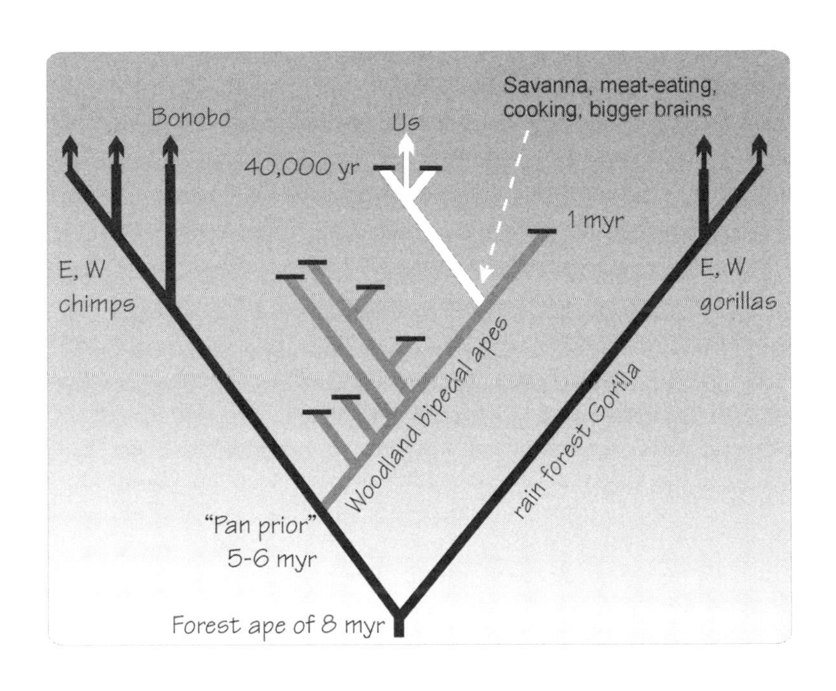
Savanna, meat-eating, cooking, bigger brains
Bonobo
Us
40,000 yr
1 myr
E, W chimps
E, W gorillas
Woodland bipedal apes
rain forest Gorilla
"Pan prior" 5-6 myr
Forest ape of 8 myr

To:	Human Evolution E-Seminar
From:	William H. Calvin
Location:	0.61215°N 36.02245°E 985m ASL
	Lake Baringo

Subject: **The earliest hominids**

I'm looking northwest from the shores of Lake Baringo toward the Tugen Hills. They have long been thought to be a good place to look for missing links, as their layers go from about 250,000 years old down to about 14 million years old. In a deep ravine near Kapsomin, about 50 km from here, a French-Kenyan paleoanthro team is excavating the five earliest-known hominid individuals. The nearby volcanic layers seem to date them back to the Late Miocene, about 6 million years ago.

It will be years yet before the paleoanthropology community agrees on how to place them, together with the new finds in the Middle Awash region of Ethiopia at 5.8 million; both early hominids appear to have lived in relatively wet woodlands. Those dates are right about when the DNA dating suggests we last shared a common ancestor with the chimp-bonobo branch. "Millennium Man" is upright but it still shows signs (paleoanthropologists love to talk about curved bones) of apelike specializations for life in the trees.

Ron Clarke's new Sterkfontein "Stw573" skeleton at 3.3 million years ago, with its weight-bearing heel bone, still has feet that can grab a tree branch. So real efficiency in bipedal stride may have taken over three million years to develop after upright stance did – maybe even as long as it took brains to finally start enlarging at 2.4 million years ago.

Doing paleoanthropology is much like doing a jigsaw puzzle without a picture of the pattern – and where you have to dig to discover the pieces! And then the pieces are worn and broken, making them ambiguous, able to fit into several different places in your imagined pattern. You guess at the pattern, using information from elsewhere in evolutionary biology and paleontology as a guide, but the general guiding principles such as Darwinism and analogies from development are very broad, leaving lots of room for sheer happenstance.

Human evolution from an apelike ancestor only happened once, and the general principles merely tell you something about what the average result might be if you ran the experiment a hundred times. And so paleoanthropologists have to guess a lot. This makes for lively controversies. Reading a historical account of paleoanthro (try Ian Tattersall's *The Fossil Trail* or Alan Walker and Pat Shipman's *The Wisdom of the Bones*) shows you how the current ideas have been built up. The track record shows you how likely it is that the current ideas are also missing a few important things.

Walking back from dinner in the dark, fully bipedal, we ran into the hotel's uniformed askari. My cousin asked him in Swahili if hippos really came up on the lawn outside our front doors. Oh yes, the guard said, pulling out his flashlight, and he led us down the path past our cabins – and there in the dim light were two full-sized hippos, mowing the grass.

They come up all through the night in different groups, sometimes a half-dozen at a time. So you sleep fitfully under your mosquito netting, listening for exaggerated treble and bass sounds, for malarial buzzing inside the room and for two-ton munching sounds just outside. At least no troop of baboons socialized on my roof for half the night.

WE TOOK A NICE HOUR-LONG BOAT RIDE in the early morning light and saw hippos chasing one another, crocs beginning to sun themselves on shore, herons hopefully inspecting the shallows, some cormorants diving farther offshore. Besides all the colorful small birds (weavers are very common here), there were hornbills, Marabou storks, and fish eagles.

People were coming down to the lake edge to haul water, wash clothes, and fish. We saw the little one-person boat of lashed-together firewood called a coracle, about the size of a floating lawn chair. Like the tidal fish trap, it doesn't hold water, but the sticks keep the fisherman afloat. His catch is kept in the bottom of the boat and benefits from the constant change of water. The fisherman paddles with split shells, one in each hand. Close inshore (where it's shallow enough to wade ashore), and at the flat water of dawn, this technique is just fine. It provides an interesting view of how boats might have gotten started and become valued.

Then after breakfast, we looked up to see a camel working over the local trees. Perhaps, I said speculatively, some camels have been driven south from Lake Turkana by the current dry spell. Around here it has been raining on and off for the last week, so we are seeing desert-bloom conditions. We almost stumbled over a medium-large tortoise crossing the grass to browse on the well-watered bushes. The hotels don't have to bait the premises to get animals to come visit the tourists. All they have to do is to water the plants and wait.

SO THE COUSIN, who is busy finishing up a book manuscript, and I were sitting in the shade, keeping an eye on the camel and the hippos from our lawn chairs, each nursing the battery life in a laptop computer while trying to get some work done. We kept hearing this faint bugle-like noise. It sounded somewhat like one of those miscellaneous computer vocalizations. But it stopped after several repetitions.

I'd already been fooled by a bird that sounded like a truck back-up warning horn, and another that sounded like a telephone. I was perfectly willing to believe in a bird-call that duplicated a computer sound, just like those starlings in Copenhagen that imitate cell-phone rings and perplex the pedestrians. A half-hour later, we heard that bugle sound again. And again. We kept commenting on it, speculating about bird mimicry, and going back to work.

Then a window popped open on my computer screen. It contained the name of the cousin's computer, not mine. Realization dawned. The infra-red ports on the two laptops had discovered each other, had exchanged names, and had been repeatedly announcing the fact. I'm sure no one else at Baringo (and certainly not the camel) understood why we laughed uproariously for minutes, at the absurdity of it all happening in this particular setting.

After lunch, we saw a fish eagle and two crested cranes near the dock. That's when the camel came over to us, looking for a handout. If a camel could ever be said to have an ingratiating smile - well, no, I suppose this one just looked opportunistic. So it is likely an experienced local camel, well versed in the ways of tourists, not a recent refugee from the wilds of Turkana.

There is something fascinating about science. One gets such wholesale returns of conjecture out of such a trifling investment of fact.
– MARK TWAIN, *Life on the Mississippi,* 1883

Some speculations can be disproved faster than others, which may take a lifetime to show their flaws. This is what makes paleoanthropology so frustrating.

To:	Human Evolution E-Seminar
From:	William H. Calvin
Location:	0.77217°S 36.42091°E 1,940m ASL Lake Naivasha

Subject: **Droughts even in good times**

Back in the Southern Hemisphere once again. We watched, I confess, the little handheld GPS unit to mark the exact (within three meters, or so it claims) crossing line, the place where there is no Coriolis effect to deflect moving bodies. No ceremony, not even toasting with water bottles, as I'm traveling with people who live on the equator.

Didn't even see the guy at the equatorial tourist trap, the one who claims to show you whirlpools reversing direction as you step across the equator. He has a basin of water with a hole in the bottom, the whirlpool spinning clockwise. When he steps across the line and refills the basin, it spins counterclockwise. This isn't, alas, a demonstration of the Coriolis effect as it is intended to be, but only an example of how people fool themselves. The Coriolis effect is vanishingly small at the equator, and for hundreds of miles around.

Furthermore, whirlpools in washbasins and bathtubs do not obey the Coriolis effect rules, even at higher latitudes. Hurricanes do, but there the Coriolis effects act for long times and over long distances. What direction a small whirlpool turns is just happenstance, as any science class would know if the teacher sent them home to survey their sinks and toilet bowls, tabulating the results from a dozen homes.

It's likely that this guy doesn't know all this; indeed, he probably isn't a knowing charlatan at all, but has just empirically learned what works to initiate a whirlpool in the right direction (how he rotates the bowl to show visitors, how he pours the water when refilling the bowl). Such things can operate at a subconscious level, and any number of scientists have gotten fooled in a similar way. The difference is that science is pretty systematic about discovering such errors and moving on.

HERE AT LAKE NAIVASHA, there are also hippos in the night (mama, papa, and junior, I was told at breakfast by the cousin, who asked the watchman to wake her up when they appeared). The watchmen here usually chase the hippos away, at least when tourists aren't awake to see them.

It seems they pull up the grass by the roots, leaving unsightly spots. The hotel planted the wrong kind of grass, probably back in the 1930s when this was an overnight stop for the "flying boats" that provided passenger air service between London and South Africa. Thus I see two grown men bent over, swinging long knives to cut the exotic grass.

Only 26 km west of here, up the Mau Escarpment, is Enkapune Ya Muto rock shelter ("Twilight Cave"), currently a hot topic because it contains the earliest evidence of beads – such decorative art is evidence of the modern mind. About 50,000 years old, it is earlier than in Europe (where cave art is the more spectacular evidence). All of those millions of years of bigger brains, and finally evidence of thinking somewhat like us.

A short boat ride away is Crescent Island, where one can walk, in the company of a guide with whom the herds are familiar, among the giraffes and waterbuck and gazelles. Obsidian flakes are everywhere, some of which are just the sort that hominid toolmakers would have prized. Some microliths can be found here, even hafted ones from a few thousand years ago. The reason that volcanic glass is so prevalent is that, just offshore, there's an old volcano lurking in the depths.

My cousin kept exclaiming over the obsidian, passing me one flake after another. I kept saying, after a brief inspection, that the proffered flake was probably not archaeological, but merely happenstance. Still, if you were in the *objet trouvé* stage of tool use, this island would have been heaven, what with such single-edged razorblades everywhere. Such a place could have been where hominids discovered the virtues of sharp edges and, when they exhausted the local supply, made the transition from found-object tool use to Glynn Isaac's shatter-and-search toolmaking.

The giraffes and the archaeologically-suggestive obsidian flakes are surely the reason why most people visit Crescent Island, but I actually came because of reading about the climate cores recently drilled offshore. Crescent Island Crater is underwater, just offshore. Old volcanic craters are not uncommon hereabouts, but the significance of this one is that it provided a protected underwater basin from which comes a lake-bottom core, one with a nice 1,100-year-long record of local climate, showing all its ups and downs via the inferred salinity of the old lake bottom layers.

THE STORY TOLD by the Crescent Island crater sediments is that the Medieval Warm Period (from about 500 to 1315) was a bad time for Africa. It is known from other sources that these were years of drought-induced famine, political unrest, and large-scale migration of tribes. What the cores say is that the lake shrank dramatically, and for many decades at a time.

Paradoxically, the Little Ice Age (roughly 1315-1865, when most of the world was generally about 1°C cooler, thanks to one of those minor 1,500-year-long climate rhythms) was a good time in East Africa, thanks to how it affected East African rainfall. The good times were relatively uneventful periods of political stability, consolidation of kingdoms, and agricultural success. And thus growth of populations.

But what interrupted even the five-century-long good times in East Africa were serious episodes of bad times. Such were concentrated in three periods: around 1390 to 1420, again from 1560 to 1625, and then from 1760 to 1840, periods when Lake Naivasha (and many a big lake in East Africa) was shrunken and salty.

Now the wind grew strong and hard, it worked at the rain crust in the corn fields. Little by little the sky was darkened by the mixing dust, and the wind felt over the earth, loosened the dust and carried it away.

– JOHN STEINBECK,
The Grapes of Wrath, 1939

So, even in the absence of human modification of climate via fossil fuels and cutting down forests, it looks as if Africa is subject to episodes of prolonged (30, 65, and 80 years) drought even in otherwise good times. And if you live elsewhere, don't feel smug about your ancestors having had the good sense to emigrate from Africa (everyone's ancestors used to live here 50,000 years ago). There have been big, prolonged droughts in North America as well, and the evidence is accumulating elsewhere. There's a lovely set of fossil tree rings from Chile, a 1229-year-long period from sometime about 50,000 years ago, which also shows droughts lasting a century, with abrupt onsets and ends.

Droughts are often regional, such as the Dust Bowl of American Midwest and Great Plains from 1931-1939. Numerous farms had to be abandoned, and overall agricultural productivity dropped sharply; my mother tells me that when it rained in Kansas City, it rained mud.

Such droughts may be simply initiated by chance events, such as random fluctuations in storm tracks over the years. But they are sustained by feedbacks that make them worse and delay recovery. A few weeks of abnormally hot weather may dry out the topsoil, reducing what plants absorb through their roots. As their leaves wilt, the ground beneath the plants becomes even hotter. As plants cease evaporating groundwater, the air becomes even less humid - and since this near-ground humidity is about half of what condenses into summer rainstorms, there is even less rain. Stressed plants may die. What rain does fall may then run off quickly, since dead plants no longer extract water from the topsoil - and so the water table drops. Once the water table drops significantly, no plants grow the next spring.

So there is a self-perpetuating aspect to a drought, once it gets started. Only another chance event - perhaps decades later - that happens to bring a lot of rain for several years in a row will manage to restart the vegetation.

Sometimes there are seesaws operating, where one region improves at the expense of another. But there are also some droughts that are worldwide, everywhere getting hit at about the same time. Everyone loses (except maybe for the waterhole predators), almost everywhere (except maybe Antarctica and the South Atlantic Ocean, where few people can live). They are the aforementioned "abrupt cooling episodes" but they could equally well be called "severe drought episodes" or "dust storm centuries." Temperature is often the easiest thing to measure, thanks to the oxygen isotope ratio correlating with air temperature, but it is not necessarily the most relevant.

> **[Our] results highlight the sensitivity of precipitation patterns to nonlinear or threshold-crossing climate change, and underscore the potential for future climate impacts on society via drought or flooding rather than temperature per se.**
> – JEFFREY P. SEVERINGHAUS & EDWARD J. BROOK, 2000

We tend to concentrate on the downside of droughts because of all the human misery they cause. But an evolutionary biologist also looks at the recovery, because the transition is often a boom time. Things become possible in boom times that are difficult in the more static periods before and after the transition period.

SUCCESSIVE WAVES OF IMMIGRANTS, who then compete and interbreed with one another, is what has happened all over the globe. It probably happened as *Homo erectus* spread out of Africa almost two million years ago. It may have happened again a half million years ago as *Homo heidelbergensis* spread out of Africa. And it surely happened again as *Homo sapiens* spread out of Africa during this last ice age.

We're accustomed to thinking about this as gradual spread, associated with gradual climate change and gradual retreats of ice sheets. Migrations are, however, stimulated by bad times, as in those three East African droughts during otherwise good times. Further, we recognize that migrations from, say, Africa to China are not Lewis-and-Clark-style expeditions with a goal in mind. Rather, they are likely a series of successive occupations along the way, with populations following their favored foods up and down in elevation over the years as things shift from warm and wet to cool and dry.

For humans, this likely meant following the grazing animals up and down, depending on the elevation that the grass grew without turning into bushlands. The temporary grass after a fire might lead them in new directions. This occasionally led them within hiking distance of a mountain pass, whereupon they discovered more prey on the other side - and so they were pumped over the pass. There is even a suggestion that such slow climatic pumping could have carried Out of Africa populations from one river valley to another along a northern "Silk Road" eastward to China.

But now that we see the role of droughts more clearly, and now that we see the Ice Ages revealed as the Chattering Ages, it is worth rethinking the pumping aspect. What additional factors might be coming into play as human evolution was given abrupt opportunities and challenges by the abrupt warmings and coolings? Slow strokes on most pumps have different yields than fast strokes, leakiness being what it is.

An abrupt drought can provide challenges (having to eat an entirely different diet because the customary prey and plants disappear) that may cause many subpopulations to fail altogether. The European settlement in Greenland, established during the medieval warming in the year 982, died out by 1540 during the Little Ice Age (though the Greenland Inuit, with their better boat-building and clothes-making technologies, survived). The Little Ice Age temperature fluctuations were only a fraction of a whiplash cooling.

Though economic and political competition has surely been with us for a long time, and warfare and genocide have provided many examples of group extinction, remember that groups need not compete against other groups like sports teams in order to evolve by group selection. Instead, the group's characteristics may cause it to thrive or fail, perhaps even unaware of other groups.

BY APE STANDARDS, we humans are quite versatile, a jack of all trades compared to the rest of the animal kingdom. We can eat vegetarian diets like the gorillas do, thanks to our modern ability to prepare a wide variety of otherwise toxic plant foods through soaking and cooking. This allows us to live in many different climates, and without a gorilla-length gut. We can also survive on very few plants, as the carnivores do, thanks to augmenting the rudimentary ape hunting skills with those of throwing and toolmaking.

There are indeed other omnivores, such as the bears and chimpanzees, but we've carried versatility to an extreme. Some of this is cultural, aided greatly by our language abilities, but much of it seems to have worked its way into the gene pool. We see year-old children hammering on their plates, a developmental program prompting a behavior that was greatly augmented in the last six million years of ape-to-human evolution. Older children come with maddening predispositions to throw rocks downhill, too.

Even in a modern vegetarian society, those hunting predispositions are there, facilitating a rediscovery of throwing techniques and their augmentation via tools such as spears. Though children growing up may lack role models for the techniques, our genes have carried along many predispositions to discover things, ones that helped our ancestors survive, once upon a time, in some different climate.

Let the climate abruptly dry and some groups will survive better than others, simply because they have the versatility to rediscover some plant- or meat-acquiring technique that had fallen out of use dozens of generations earlier. The flickering climate has been worse than the "bait and switch" schemes of the disreputable advertisers: offer a "come on" in the form of a suddenly blossoming environmental niche (those abrupt warmings from the cool-and-dry mode of climate), and then, once the population has grown to fill the space available, change the name of the game with a sudden reversion to the cool-and-dry-and-windy-and-dusty mode - a switch so sudden and so serious that it forces a scramble for survival in an increasingly isolated subpopulation, and all within a single generation.

THE FRAGMENTATION OF POPULATIONS, and the rediscovery of empty niches, are among the major accelerators of Darwinian evolution. Many animal species were likely affected by such expand-and-shrink cycles, not just our particular ancestors. Some group selection via climate has probably affected other social species, such as the dogs, using the same happenstance makeup of small subpopulations and their survival via social cooperation.

A round man cannot be expected to fit into a square hole right away. He must have time to modify his shape.
– MARK TWAIN (1835-1910)

So far as I know, bears and chimpanzees haven't had a brain boom during the last several million years. Something must have been special about how our omnivorous ancestors made their living. Something engaged the gearshift which allowed climate-driven fragmentation to help ratchet up our cherished beyond-the-apes attributes - altruism, simple language, structured thought and language, and an ability to anticipate the outcome of a proposed course of action, what it takes to achieve ethical behavior (or plan a war). What was this something?

As I mentioned earlier, the brain doesn't evolve via a bump here and a bump there, as different functions come under natural selection. Except for smell, it looks as if it's something closer to "enlarge one, enlarge them all" - not exactly the mosaic selection which adaptationist reasoning tends to assume.

Furthermore, despite the names we tend to give cortical areas like "visual association cortex," many cortical areas are multifunctional. This means that you don't need natural selection for music, for example, to evolve musical abilities; they could easily be a spare-time use of the language machinery of the brain, with its ability to handle structure. Music could have been bootstrapped by language.

And, as handy as structured language is, it could have gotten its initial big boost up from ape-level utterances via a similar route: a spare-time use of neural circuits that were being shaped by natural selection for some other function with a more immediate payoff than intellect - in other words, the payoff that augments the neural circuits need not be payoffs from the higher intellectual functions themselves. Indeed, the language areas of the present-day human brain have a lot of overlap with the areas important for hand-arm movements. Were throwing and hammering particularly useful for making a living, then there might have been some spare-time use of the same circuitry in the evenings for gossip. The growth of language abilities might have, in the initial stages, simply been a free, spare-time consequence of natural selection for hunting skills. Language could have been bootstrapped by precision ballistic skills, though eventually paying for further improvements via its own virtues.

Since I proposed that in 1981, another such carryover with bootstrapping potential has been suggested. In our book *Lingua ex Machina*, the linguist Derek Bickerton points out that the evolution of altruism could have provided the mental categories needed for argument structure in syntax. Free-loading is the big problem with sharing; everyone loves a freebie, a social *objet trouvé.* Happenstance clustering of individuals likely to share helps to overcome this evolutionary problem. Another aid is to share mostly with individuals themselves likely to share. But how do you know an individual is a sharer? It would be handy, as Richard Dawkins once noted, if they all wore red beards or some other mark that distinguished them from the habitual cheaters.

One strategy is to share initially, but keep track of reciprocity, refusing to share on some subsequent occasion if repeatedly disappointed. "But you owe me!" requires some mental categories for debt. The task of remembering who owes what to whom is amazingly like linguistic argument structure (those word categories involving actors, recipients, beneficiaries, and so forth), which provide major clues to understanding a story-like sentence about who did what to whom. The same categories that are so handy for minimizing free-loading in sharing are very similar to those needed for fancy sentences. Language structure could have been bootstrapped by the categories needed for altruism to succeed.

If our ancestors already had protolanguage (words and short sentences like two-year-olds), then it isn't a big step up to structured speech or sign, given you'd already have a language-ready listener preadapted with the mental categories to parse your structuring into phrases and clauses.

Fragmentation into shrinking subpopulations, thanks to an abrupt cooling, would place a lot of importance on versatility in food finding. And the happenstance clustering of small groups selected from a larger population would have occasionally created groups where altruism had enough practitioners to make a difference in the crunch. Language, too, needs a critical mass if children are to be exposed to sufficient examples, so that they pick up enough by imitation to be useful later.

That "something," which made abrupt climate changes different for our ancestors than for the other omnivores, isn't really a settled scientific question. But it may well have to do with the tools that our ancestors invented: the action-at-a-distance of projectile predation, the sharp tools needed for food preparation, and the "debt tools" of altruism. Other things built upon them, such as the wonderful toolkit that we call our vocabulary, such as our abilities to speculate about the future and engage in beyond-the-apes levels of social manipulation. But the basics are exactly what might make a big difference in subpopulation survival during the fragmenting population crashes – and they are things that the other omnivores haven't also invented.

I HAVE BEEN CASTING SOME DOUBT on the traditional Darwinian interpretation of gradual "progress," not because I think it is incorrect but only because I think that, unaided, it is usually too slow and too easily reversed. First I said that running in place – "automatic gradualism" – is pretty slow. Second, that the expand-and-contract population cycles also might not accomplish much in the way of adaptations. These everyday and every-century aspects might be minor in comparison to periods that do most of the "evolutionary work."

Then I said that it was when even refugia came under pressure that adaptations really mattered. Furthermore, it may take a lucky combination of adaptation and speciation to keep an adaptation from backsliding when immigrants arrive (and dilution of the "progress" threatens), and small groups are a better setup for speciation as well.

Now let me revisit the problem of repeating the course for additional credit – some improvements have growth curves, and others do not. As a generation of anthropologists emphasized, back when Man the Hunter was out of fashion and Woman the Gatherer was being emphasized, the carrying bag must have been a very important invention for both gathering and for small-game hunting. I agree. Yet one cannot reinvent the carrying bag for extra credit. Fortunately, a few types of invention can be repeated, much as the occasional college course (say, undergraduate research or music technique courses) can be repeated for additional credit.

The standard example is that many aquatic mammals have discovered that a small reduction in body hair buys them greater swimming efficiency. Another reduction buys them even more. No matter where along this "growth curve" they are, another increment has additional rewards. Yet you can only become so naked. Some growth curves plateau.

Some growth curves are also steeper than others, faster at driving evolution than other candidates. There are at least two aspects of the abrupt boom-and-bust scenario that have long growth curves. They involve things from that chunnel list, where we humans have considerably enhanced abilities over the great apes.

The side-of-the-barn accuracy needed for flinging branches into waterhole herds doesn't have much of a growth curve by itself (it doesn't matter which one you trip up), making it more like the carrying bag. Important, but you need something entirely different for your next act. Still, it could have gotten hunters on to the bottom on the precision-throwing growth curve.

Being able to hit smaller herds has an even more frequent payoff. Once herds become wary, then an ability to hit the target from a greater distance becomes important. This has the incidental benefit of reducing risk to the hunter. No matter how many times you successfully double your throwing distance, it has an additional payoff, with no plateau in sight. Eventually you can become accurate enough to hit lone animals from the distances achieved by modern baseball pitchers. Important technological inventions improve throwing further, such as spears and launching sticks. Note that each improvement confers an additional payoff:

additional days that you and your offspring can eat high-calorie meat.

Sharing has a similarly long growth curve. You can share more things, over longer periods of time, with more people, and so forth. Yet there are few examples in the primates, except for mothers sharing food with their offspring. There is one stunning exception to this mothers-only rule: fresh meat is shared by chimpanzees, and not just with the other chimps who took part in the chase. The palm-out begging gesture is directed toward the possessor of the meat, even if lower in the usual dominance hierarchy (were it fruit, high-ranking animals would likely plunder the food). Almost everyone that is persistent gets a scrap of raw meat. No theorist of social behavior would have dared to hypothesize such an exception for meat, for fear of being laughed off the stage - yet there it is, in one field study after another.

Glynn Isaac suggested a quarter-century ago that meat may have provided a social currency, and various studies since then have suggested that meat cannot be understood solely in terms of calories contributed to the diet. My suggestion is that hunting is crucial mostly for surviving the abrupt droughts, where meat on the hoof is the major remaining resource - but that the social (and sexual selection) value of meat is what sustains the cultural aspects of hunting in between the stressful occasions - and that the boom time aftermaths allow the hunting-capable population to expand faster than otherwise.

If carnivory was indeed the catalyst for the evolution of sharing, it is hard to escape the conclusion that human morality is steeped in animal blood. When we give money to begging strangers, ship food to starving people, or vote for measures that benefit the poor, we follow impulses shaped since the time our ancestors began to cluster around meat possessors. At the center of the original circle we find a prize hard to get but desired by many. . . . This small, sympathetic circle grew steadily to encompass all of humanity - if not in practice then at least in principle. . . . Given the circle's proposed origin, it is profoundly ironic that its expansion should culminate in a plea for vegetarianism.

– FRANS DE WAAL,

Good Natured. The Origins of Right and Wrong, 1996

To:	Human Evolution E-Seminar
From:	William H. Calvin
Location:	2.99599°S 35.35348°E 1480m ASL Olduvai Gorge

Subject: **Degrees of Separation**

Though the Taung child's skull and brain endocast was the first fossil hominid of distinction, its discovery in 1924 was not heralded by *National Geographic* photographers and television reporters. It wasn't in a stone tool context, and it was ambiguous because the juvenile skulls of closely related species look so alike. One wanted a better-differentiated adult to see what it really looked like, an ape or something more human.

To most people who don't know the Taung history, Olduvai Gorge is where the first hominid skull was discovered in 1959. That's certainly what I thought because later in that year, when Louis Leakey was showing off their finds in the U.S., he came to visit the graduate anthropology seminar that I was taking. Out of his pocket, he pulled some stone tools and several absurdly large teeth and handed them to me, one by one. They were probably only casts of the true fossil teeth, of course, but I couldn't tell – I was a physics undergraduate who had talked myself into Melville Herskovits's famous course without ever having taken the introductory anthro courses. "Man the Toolmaker" was the motto back then, and as we handled the evidence, we caught Leakey's enthusiasm for finding a place where some of the pieces might fit together, in the walls of Olduvai Gorge.

AS YOU DRIVE WEST DOWN FROM NGORONGORO in the crater highlands, you come to Olduvai, a gorge with layers exposed that go back 2 million years. Higher up its walls are lots of handaxes. No wonder Louis and Mary Leakey were attracted to Olduvai. They reasoned that the stone tools they found here made it likely that they would eventually find some of the bones of the toolmakers. And after several decades, they found a skull here in the bottom layers, one with such large teeth that he was sometimes called Nutcracker Man. Ironically, Zinj now turns out to be one of the hyper-robust australopiths, and thus not so likely to have been one of the better toolmakers. The second irony is that the vindicated Taung looks more like us than Zinj. Indeed, with its concave face, Zinj is now considered one of the strangest-looking hominid skulls of all.

A lot has been learned since 1959 at Olduvai. While it bottoms out at about 2 million years, Laetoli, 45 km to the south, has earlier layers – including the volcanic ash fall that preserved 3.56-million-year-old footprints of several bipedal apes. I was happy to see the Gorge Museum and the Zinj discovery site, especially after discovering how few tourists in Kenya seem to be interested in the paleoanthropology; our very experienced Kenyan guide had never had occasion to go to either Olorgesailie or Kariandusi. The tourist booksellers confirmed my suspicion: big game and birds rank far higher for tourists than some of the world's very best sites for showing the epic story of becoming human. If Olduvai were not on the road from Ngorongoro to the Serengeti, I wonder how many people would go out of their way to visit it. But then half of Americans still think that humans lived at the same time as dinosaurs, a little error of 65 million years.

JUST SOUTHEAST OF NGORONGORO is Lake Manyara, a modern lake much like "Lake Olduvai" used to be when Zinj was around. It's in the rift valley and, as you drive down the brushy escarpment, passing a variety of baobab trees, you see the complete range of hominid habitats during evolutionary time as you near the lake.

Near the springs where groundwater from the crater highlands emerges from the hillside, you see forest. Thanks to reliable shade in this forested band, the direct sunlight in the dry season does not readily bake all of the moisture out of the soil.

Downhill, with less water, the trees thin out. Aided by the elephants that knock down trees and create clearings, there is an open woodland where giraffes browse the treetops.

Finally, there is a savanna of grass near the lake, with only the occasional tree for shade or refuge.

Apes mostly live in forests (and occasional swamps). Before about 1.8 million years ago, hominids lived in the open woodlands, not the savannas. They probably dug out a lot of roots and potatolike "underground storage organs" not found in forests; in woodlands, their moisture is needed to get through the dry season.

It was only later that hominids made their living out on the savanna with regularity. And they probably weren't eating grass.

JUST SOUTHWEST OF NGORONGORO is where the Hadza live, downhill around Lake Eyasi.

The Hadza are still a hunting and gathering people, one of the last ones left. They have no system of chiefs, no permanent villages. They roam the swamps and forests hunting small game (mostly alone, though they may occasionally cooperate when tackling a whole troop of baboons at once), gathering roots and tubers, living in small camps of a dozen or two and moving on every week or so.

They speak a click language like that of the Kalahari San people a half-continent to the south, have a small pygmy-like stature despite a good diet, but look somewhat like the Arabs who spread down the coast via Somalia. It is hoped that their DNA will help clarify their roots.

They are surrounded by cattle herders and fledgling agriculturalists of other tribes, speaking very different languages, people who are busy burning the forests to clear the land for grass or crops. The Tanzanian government, a few decades ago, tried to give the Hadza some cleared land, livestock, grain, and tools - but in less than a generation's time they fled back to their traditional swamps and forests. Soon the Hadza's traditional way of life, thanks to their progressive neighbors, will vanish, leaving them without heritage or hope.

The Hadza give us a rare view of how life can be lived without much planning or agriculture, without chiefs, without a settled existence. While there are a few degrees of cultural separation between the Hadza and the Tanzanian city-dwellers, the Hadza are behaviorally modern: their art is body decoration and perhaps some cave paintings, their language is elaborate, and they play games with rules.

We can try to imagine the Hadza without modernity, sitting around skillfully-lit campfires pounding roots and sharpening stones, perhaps speaking the short sentences of a modern two-year-old - but without storytelling, gossip, and gambling, without showing much concern for the past or future. And in doing so, we can perhaps approximate what *Homo sapiens* was doing with a modern-sized brain in Africa in the 100,000 years before Africans invented (or evolved) behavioral modernity and spread into Eurasia. More later.

AFRICANS WHO GET GOOD EDUCATIONS and are encountered in their roles as pilots, professors, and neurosurgeons show that the biological basis is there for doing the things that the Out of Africa peoples do so well, around the rest of the world. Within Africa, raised one way without much western-style education and then forced to deal with a technological culture in order to support their families, they may not show those strengths, for all the reasons that Peter Mattheissen wrote about in *The Tree Where Man Was Born* as he described the old colonial days:

> It is often said that Africans cannot lay straight paths or plow straight furrows, screw bottle caps, use rifle sights - in short, that they have no sense of geometric order, much less time, since there is nothing of this sort in nature to instruct them. "Have to watch these chaps every minute," white East Africans will tell you. "Can't do the first thing for themselves." But perhaps the proper word is "won't." Most Africans are so accustomed to having decisions made for them..., and to carrying out instructions to the letter to avoid abuse, that only rarely do they think out what they are doing, much less take initiative. Rather, they move dully through dull menial tasks - working automatically, without thought, may be all that makes such labor bearable - preferring to be thought stupid than to get in trouble, and at the same time gleeful when calamities occur. Stubborn, apathetic, and perverse, they observe the letter of their instructions, not the spirit of them . . . . Even among white East Africans and black who converse easily in English or Swahili, the problem appears to be mutual boredom, which comes about because both find the interests of the other trivial, and their ideas therefore of small consequence.

Dated though this is, I bring it up because I worry that much of the world could become like that, a world where most of us will not have the education to know how things work - that we will become trapped in a world where you have to wait for someone knowledgeable to tell you what to do next and are thus unable to undertake initiatives on your own.

One of the shocks, in talking to Kenyans, is discovering that basic education is not a given, that parents must pay to send a child to grade school. And, being so labor intensive, the good schools cost dearly. Elsewhere with free public schools, an expensive private school may be nonetheless considered the only way for children to be properly prepared for a fast-paced world. And so the rich get richer, and the poor get poorer.

It takes a heavy commitment to quality education for all to avoid that stratification of society, those needless degrees of separation. But even the present-day United States has lost what commitment it used to have to free education of high quality. Anyone reading the annual surveys of science literacy (another example: fewer than half of Americans know that the earth orbits the sun once a year) has to wonder how badly most people are going to be left behind, further along into the 21st Century, whether they too will become "stubborn, apathetic, and perverse" toward a scientific and technological world they must view as magical, beyond their comprehension, accessible only via the right incantations.

THERE ARE MANY REASONS why public schools are neglected. The one I find most troubling is a curious aspect of practical genetics, one that an evolutionary viewpoint helps illuminate because it is nepotism at one remove, with all the same long-run disadvantages for the community at large.

Take two high-IQ parents, say 130 each, and consider their offspring. While it is true that a child of 150 IQ is more likely to come from such parents, it is also true that their average child may be only 120. (This statistical law is called regression toward the mean; it also works to bring up children of lower-IQ parents closer to the 100 average.) This doesn't apply to particular families, of course, who don't have a hundred offspring to test the probabilities, but it should collectively apply to the group of parents who send their kids to expensive schools: on average, their kids will be less amply endowed.

This is the familiar problem with family businesses and lines of royalty: the second and third generations often aren't as smart as the founders. More intensive education can, of course, help make up for any performance differences between parents and offspring. In this odd, unexpected way, high IQ parents feel more of a need for private schools for their offspring than do average parents, because they want their children to do at least as well as they themselves.

The problem is that there is another way of helping the less-well-endowed offspring of those who have made it: decreasing the competition. It helps explain the generation-ago puzzle of the mother who was against admitting women to medical school; she thereby helped to double the chances of her on-the-edge son getting admitted. If there are spaces for only one percent of the population, eliminating females meant that twice as many males made it in. And who can be sure that their child is in the top one percent?

Discrimination to reduce the competition works, alas, for any sizeable group. And while one naturally thinks of the twentieth century racial-religious examples in higher education, it also works with the following group: those who cannot afford private education, but have the talent to compete with those who can. By slowing down the competition, those well-off (and often influential) parents can help their kids to get ahead. An easy effortless way of hobbling the competition is to neglect public education.

This is, of course, not in hardly anyone's conscious reasoning – it's not intentional so much as conveniently incidental. Consciously, it will be all about their admiration for high quality (though not a parent, I regularly visit the private school in Seattle where Paul Allen and Bill Gates went to high school in the 1970s, and it is wonderful to see the educational job done really well, an example for all to emulate). Perhaps the influential parents will think that they can encourage public school quality via "competition in education" or "accountability" or "testing teachers." But good intentions are a tricky thing; when the free schools are hamstrung with regulations and underfunded, the holding-them-back result is much the same as overt discrimination against the group that cannot afford private educations. You simply get needless stratification.

If you think quality public education is expensive, consider the costs of ignorance and polarization, of an intelligent underclass that becomes stubborn, apathetic, and perverse.

To:	Human Evolution E-Seminar
From:	William H. Calvin
Location:	1.28063°S 35.08497°E 1,570m ASL Maasai Mara

Subject: **The Crash-Boom-Boom cycle**

There's nothing like the view from the edge of a watering hole. Our tents are on a slight rise, and the view consists of several elephants wading knee-deep, hosing themselves down and flapping their ears in the afternoon heat. Two wart hogs wander among our tents in an accustomed manner, showing no interest in anything except grass - which they eat while kneeling on their front legs. This shows a proper reverence for grass.

The baboons are not reverent, not about anything. Should they prove obnoxious, the Maasai guard is armed with two short bare sticks looking like thick drumsticks, which he bangs together to threaten them. To the baboons, the askari's sticks must sound like giant canine teeth, gnashing in threat. Perhaps after a second beer, I might speculate about how rattles and drumsticks came to be invented by our ancestors, but not so early in the day.

Rivers tend to meander if they aren't hurrying rapidly downhill. Since meanders are often eliminated when a flood erodes through a bank to create a more direct route, there are lots of dead-end channels ("abandoned meanders") which make excellent watering spots. And they tend to fill up with delicious vegetation, something our upright ancestors might have noticed as well. I can even imagine them as refuges from predators. If you splash water toward an animal's face, it stirs up the blink reflex and tends to make them cautious about proceeding. Offshore islands with trees are a good refuge.

Gallery woods may form along the river, trees extending as far from the water as roots will reach, and our ancestors might have found them comforting. Could we somehow test australopithecines on Gordon Orians's landscape esthetics, they might well prefer the combination of shoreline features and forest features - not isolated trees in a grassland with a pond, what we seem to prefer.

The Maasai Mara is the northern end of the Serengeti Plain, grasslands interrupted by the occasional acacia tree and the even more occasional waterhole or stream. With the aid of some DNA-coded recipes, the grass is readily turned into herds of elephant, zebra, bovids of various types, and antelope of even more types. The various seeds feed the baboons. The trees and bushes feed the giraffes and other browsers. The grazers in particular feed the lions, cheetah, and leopards. The leftovers feed the jackals, hyenas, and vultures (though they too will hunt on occasions).

As Phillip Tobias said, "From 1925 to 1995 almost everyone grew up on the 'received wisdom' that the *Hominidae* (the family of mankind) was born on the savannah, believed to have been the ideal crucible in which the strange form of locomotion known as bipedalism came into being." But since Serengeti-scale savanna scenes are only one or two million years old, our earliest after-the-apes ancestors didn't move into this scene so much as they evolved with it, as the slower climate changes and uplift produced more grass and less forest.

Even worse for the savanna bipedalism hypothesis, the earliest bipedal hominid fossils are found with flora and fauna that are associated with woodlands. So before *Homo erectus* days (they are found in much more arid areas), hominid habitat may have been forest fringes - but perhaps ones near a river or water hole. Both lowland gorillas and bonobos also like swampy areas.

For all three of the major steps up to becoming human - upright posture, bigger brains, and the modern mind (a separate problem, more in a minute) – we are back to square one, having eliminated the favorite ideas of prior decades through a lot of hard work by the primatologists, paleoanthropologists, and archaeologists. Not one of those steps now has a fleshed out, widely agreed upon theory for why it happened, though there are some interesting candidates. This is progress, but we badly need fresh ideas.

FORESTS ARE PARTICULARLY VULNERABLE TO FIRE, because they can spread it from tree to tree so easily. And it can take centuries for a forest to come back, usually through a succession of grass to brush to small trees and eventually to climax-forest trees. When brush burns, you may get grass for years afterward.

(No one knows how far back the practice goes of setting fires to clear the underbrush, improving the forage for antelope the next year, and favoring the various plants that produce berries, but such fires are a common practice of preagricultural peoples around the world. When flying over agricultural parts of Kansas, one can still see mile-long fires as fields are burned on a windless day, with a vertical curtain of smoke hanging in the still air. There is a great deal of forest and field burning in India and southeast Asia today.)

When a drought hits, the lake margins become refugia for various species. Many forests then burn down, once lightning discovers dry trees. A year later, the former forests become temporary savannas. This creates a boom time for the grazing animals, with their short generation times (no "childhood" at all – many females become pregnant shortly after they themselves are weaned), allowing population doubling and redoubling in just a few years.

So the aftermath of a drought, one severe enough to burn off a lot of forest, is potentially a boom time for experienced predators on grazing animals. Which our ancestors likely were, and probably well before two million years ago.

MORNING AT THE WATERHOLE features birds that sound like oboes. Now there is a family of four wart hogs mowing the grass, not the same ones as yesterday. I see a fish eagle perched, eaglelike, in a crooked treetop with a good view. There is a special quality of low-angle morning sunlight, casting features in high relief compared to the flat lighting of midday. Perhaps the eagle has also noticed.

What I need is some equivalent of early morning sunlight, for throwing into sharp relief the mental traits that evolved alongside, or ahead of, the stones and bones we can actually recover. To appreciate the importance of our abilities to reflect and speculate, you have to have a foundation in the mindless stuff like feelings and emotions - among which are landscape esthetics.

To appreciate the occasional emergent properties, and the new uses for old things, you need to get behind the names we give things. (Is there really "language cortex" or have useful-for-language areas kept their original functions for the most part? A "gene for symbolic behavior," or just another application of a more general behavioral trait?)

The physiologist Albert Szent-Gyorgyi liked to say that, for every complex problem, there is a simple, easy to understand, incorrect answer. We humans are very much into storytelling, and it is likely that there are some gut-level mindless criteria which make us like one story better than another (though our rational facilities may sometimes allow us to supercede the gut-level guides). We can be too readily satisfied when selected parts all seem to hang together nicely. We may, for example, argue in a circle without realizing it, making no progress at all except to satisfy our esthetic sense about what constitutes a good story. Scientists, in the course of our education and researches, encounter a lot of stories that no longer satisfy because we have learned that we (or our predecessors) were earlier fooled.

With this caution, let me talk about acquisitiveness. Birds are acquisitive of nesting materials. We tend to connect that with rearing offspring and give it a purposeful label. Packrats collect all sorts of things for building "houses." They cut branches, cactus pads, and leaves. Then they pick up anything loose to add to the structure - bottles, cans, mule droppings, bones, papers, or even mouse-traps - just to fill space. We can still ascribe purpose to this, but, because the class of objects is becoming more general, we have to be careful of our labels.

From this, we move to the acquisitiveness of the local baboons, the ones that conduct raids on our tents near the waterhole. They are reported to have a particular fondness for toothpaste, but their tastes are nonspecific enough that they are just as likely to make off with your camera bag, without checking first for anything edible. Back home in Seattle, my neighborhood is plagued by someone who is so acquisitive of battered old cars that he has left dozens of them parked on our streets, coming around occasionally to push them a little farther down the block.

Acquisitiveness is also the concept which I have applied (it's somewhere in *Lingua ex Machina*) to how the human infant builds a language machine by listening for patterns. As in nest building, there may be some general principles behind it - yet ones so loose that they don't specify too many of the particulars. First, infants begin forming up categories for the common speech sounds they hear, not whole words so much as the little units we call phonemes, less than a tenth of a second in duration. Categories allow them to generalize across speakers, so that the mother's /ba/ sound and the father's somewhat deeper /ba/ sound are treated the same despite their differences. By about a year of age, babies stop hearing many of these differences, having standardized them: a /ba/ is a /ba/ is a /ba/.

The bird that sounds like a musical telephone bell is back, causing reflex jerks of my shoulders now and then. I am trying to persuade myself that, because the call doesn't repeat at an exact interval like a true telephone bell, it isn't civilization intruding. But categorical perception keeps winning, conforming the sound to my telephone-ring standard and grabbing my attention away from other things.

By a year of age, babies are discovering patterns in the strings of phonemes and acquiring a few new words every day, just from the examples they hear (long before they begin speaking them). You can say kids are like sponges soaking up words but that's too passive a notion, one of the reasons I prefer the more active "acquisitive" as the characterization. I suspect they're searching for patterns in their environment. Eventually the kids incorporate the more reliable guesses about underlying pattern as new mental categories.

So kids have pyramided words atop the phonemes, and now have compound structures made from building blocks. But then they do it again, discovering patterns in the strings of words they hear and inferring the grammar of that particular language: ways of making plurals and past tenses and nested phrases. This happens between the ages of 18 and 36 months. Then they're off detecting patterns on even longer time scales, that of the collection of sentences we call a story. They infer that a proper story has a beginning, middle, and a wrap-up ending.

Now you can talk about this as separate instincts for phonemes, words, syntax, and narrative – or you can try out an overarching principle. Mine is that it is just repeated instances of pattern-finding acquisitiveness, on longer and longer time scales. It may, of course, turn out to be something of both the general and the specific, but what's interesting is how much you can build by just pyramiding 'acquisitiveness' concepts four times over. It's a pyramid, about like compounding atoms into molecules, molecules into crystals, and then making sharp tools out of the crystals.

ACQUISITIVENESS BY KIDS is why there might be more than just cultural spread involved when converting to widespread syntax, sometime before 50,000 years ago (more later). Yes, I can see acquisitiveness for words per se having been around for a million years or more. But pyramiding to syntax and then narratives in the preschool years, well before much plan-ahead or accurate throwing develops, happens so reliably in most modern kids that it makes me wonder if that higher-order acquisitiveness is an additional adaptation, backstopped by a speciation-like event.

Of course, it need not be true speciation but simply assortative mating. Though exceptions abound, tall people tend to select other tall people for mates, blacks tend to marry blacks, and so on. It takes little imagination to suppose that articulate women, operating in a system that allows female mate choice, would prefer equally articulate men. And so the rich get richer and the poor get poorer, once again.

Unlike true speciation, assortative mating isn't back-stopped, with a ratchet to prevent backsliding. Like dog breeds backsliding to mongrels, things can still become average. Could syntax be lost if something like prolonged suspension of female mate selection (say, via a return to harems) removed this tendency towards divergent selection? Perhaps not, because kids now have such strong multistage acquisitiveness for language that they can reinvent syntax, if no longer exposed to structured speech. Those who acquire words early enough and fast enough (nine new words every day) might, like the deaf playmates of hearing parents, invent their own structuring.

THE EARLY-MORNING LIGHT has gone, the sun is hidden behind a cloud, and I can smell a proper English breakfast, not one of your out-of-the-box breakfasts. "Out of the box" thinking is another matter, deriving from "boxed in" and the escape there from. Escaping isn't easy, and for some very fundamental reasons such as the simple categories we fall into, ones where their efficiencies for some uses serve to blind us in other ways.

After a year or so of exposure to the Japanese phoneme that lies between the English L and R sounds, for example, an infant will no longer hear the difference between L and R if not early exposed to English speakers. The infant has created an efficient box that, by standardizing variants, enables rapid processing of speech sounds. But it does mean the English words, rice and lice, now sound the same.

Well, we do exactly the same thing at higher levels, too – you see the problem in medicine all the time, where putting things in boxes serves to promote speed in decision making but also blinds the practitioners to the subtleties. One expects that it even applies to those who learn some genetics, but not enough to be able to "think out of the box" on the occasions when mutations and chromosomal rearrangements just aren't the name of the game.

Certainly most of us who contemplate climate and human evolution have gotten trapped, at one time or another, in the Ice Age Box. That name, "ice age," traps us into thinking that the last 2.5 million years were all about giant ice mountains coming and going. It traps us into thinking that such periods were icy cold when, here in the tropics, they were merely 3°C to 5°C cooler, not exactly fur coat weather. It distracts us from characterizing periods as "dry" by focusing our attention on temperature per se.

[When environments change], they usually do so pretty rapidly, at rates with which adaptation by natural selection would be hard put to keep up. When such change occurs, the quality of your adaptation to your old habitat is irrelevant, and any competitive advantage you might have had may be eliminated at a stroke.
–IAN TATTERSALL,
Becoming Human, 1998

And the "glacial" pace of the ice ages, where it takes major ice sheets 100,000 years to come and go, may well focus us on the wrong time scale. It's the things that happen in a few years, like droughts and fires, that might be the relevant aspect of the "Ice Ages" for big brains and toolmaking. Indeed it may be the decade-long transitions between states, not the steady states (warm-wet and cool-dry) themselves, that create the most relevant challenges and the most relevant new niche opportunities.

MOST OF THE EXPLANATIONS offered for prehuman evolution, as I mentioned earlier, suffer from the "Why us?" problem. An explanation needs to explain why all the changes happened to our ancestors and not to other omnivores such as the chimpanzees and the bears. My explanation, whatever its faults, at least handles that one.

The opportunities associated with bust-then-boom episodes were for a savanna-adapted ape. As apes rather than big cats, they had the brains and hands suitable for using tools. As waterhole predators they would survive the drought aspect better, with the larger lakes serving as refugia. As savanna specialists, they would have a population boom in the next generation as the grazers expanded their range into the former forests. Then dozens of generations later, when the sudden switch to warmer-and-wetter allowed a quick grasslands expansion into former deserts, there would have been a second boom time. So - at least if you eat grass or grazing animals - the cycle is two ups for each downer: Crash-Boom-Boom.

A boom time gets rare variants surviving long enough to reproduce. Nothing like this one-two opportunity happened to the chimps and bonobos, whose populations probably stayed small during the cool-and-dry phase, only slowly expanding when warm-and-wet finally came back. Note that the contract-and-boom scenario repeats for hundreds of major episodes and thousands of minor ones, thanks to the climate instabilities and their rapid transitions. Gradual changes would accumulate from repeated catastrophes. "Catastrophic gradualism," pumped by climate instability, affords us one candidate for how we ascended.

Alas, boom-and-bust also illustrates how we could lose it. I'll save that for the long flight home.

SUNSET AT THE WATERHOLE seems promising, with its long shadows and renewed breezes, but we have been sobered by the prospect of leaving this Serengeti paradise tomorrow. I learned a lot yesterday about the differences between lakes and the more ephemeral waterholes which dot the landscape here. And it provides one way of summarizing what happened in the last five million years since hominids were still close to the apes.

The big lakes that we saw up north in the Rift Valley, like Baringo and Nakuru and Naivasha, are clearly great magnets for wildlife. They vary in size over the decades but seldom actually turn salty and then dry up. One can imagine our early post-chimp ancestors hanging out around a lake, profiting from the way it focuses resources for predators. Lakes bordered by forest, with a zone of former mud flat for grass, would have made for a nice transition, the comforting landscape esthetics of the chimplike forest still close at hand.

Rivers too are always there, though some seasonally dry up enough so that you have to dig a little, to find the water flowing under the surface. Rivers often have gallery woods bordering them, again a comfort and a refuge. Australopithecines likely nested in trees at night, just to get away from the big cats. Clearly rivers are prime expansion territory in good times, when lots of kids manage to grow up at the lake edge and need territory of their own. When a drought hits later, the hominid population might have shrunken to what the lake margins could support.

Rivers have waterholes, too, in those abandoned meanders that begin to silt up. But the Serengeti also has a different type of waterhole, far from any obvious stream. It is as if there were underground streams that had had a section of their roof cave in, probably with a little help from the elephants who like to dig for mud, perhaps because their low-frequency hearing enables them to hear flowing water beneath their feet. (The Maasai cattle herders outside the park like to have elephants around, because they dig waterholes that the cattle can share, but their wildlife enthusiasm doesn't extend to competitors for grass like zebras or to predators like lions.)

The Serengeti has a foundation of crystalline rock that the ground water doesn't penetrate very well, topped by volcanics that make good soils. Indeed the elephant behavior provides many species with access to water at sites far from lakes and rivers. The savanna is dotted with lone acacia trees and lone termite mounds and, less often, with unexpected waterholes. Looking out across the savanna, you may not realize that a waterhole is there (many have no bushes or trees nearby) until several zebras climb out of it – the way horses climb up out of a ha-ha, dug by an English landscape architect to hide a fence. The waterhole really is a hole.

Landscape esthetics suggest that our ancestors made the transition to being comfortable in such settings, not merely tolerating them. And this is in spite of the fact that open savanna is prime predator territory, where there are few places to hide. Savanna away from permanent water was the great expansion opportunity for hominids, tenfold greater numbers of grazing animals clustering around those isolated waterholes. The waterholes tend to be ephemeral: sometimes watery, sometimes muddy, and sometimes really dry. This change can be seasonal but even your better waterholes can dry up for years.

Just as the lakes-to-rivers expansion would reverse in droughts to center on lakes, so the further rivers-to-treeless-waterhole expansion would have been even more vulnerable in droughts. Surely there was a lot of friction between groups and, when the drought crunch came, the peripheral savanna specialists probably tried to displace the ones that had comfortably stayed nearer the permanent water. Expand-and-contract has a bloody side to it.

Decades-long droughts, as I mentioned at Lake Naivasha, are rather common events, even in the good-times centuries of the historical record. And the ice cores now show that worse droughts have happened *worldwide* every few thousand years when warm-and-wet flips into the cool-and-dry mode (and the Amazon flow is cut in half). These droughts are far more severe and far longer, and sometimes they come in a series, a century when the climate goes mad with chattering whiplashes.

The modern human population may not be limited by predators, but we're still subject to population crashes from various causes. The Four Horsemen of the Apocalypse supplemented death with warfare, pestilence, and famine. The plague years in Europe were bad enough, where more than a third of the population died, but the natives of the Americas suffered far worse from the smallpox and tuberculosis brought by the somewhat resistant European settlers. Most of the native population was lost simply from infectious disease, running ahead of the settlers.

Famine from abrupt climate shifts has been equally dramatic and even faster, with some isolated populations wiped out entirely on many occasions. What the population biologists call "die backs" were surely a common occurrence in human evolution. The whiplash climate changes would have been far more stressful than the usual droughts, simply because of their widespread alterations in the availability of plants and prey animals. Everywhere that humans lived, climate would likely have changed simultaneously - even the tropical regions would have had changes in rainfall patterns, disrupting the food supply. The tropics may not have cooled more than 5°C, but the rainfall patterns there would have abruptly shifted within the time scan of a single generation, creating patchy resource distributions, and raising questions of refugia. Many subpopulations would not have been able to relocate before starving.

THE MODERN HUMAN POPULATION expanded from millions to billions after agriculture revolutionized things. In the United States and Canada, fewer than two percent of the population can now feed the rest, given modern technology, credit, and transportation (only a century or two ago, 80 percent were engaged in just raising food).

But the population expansion does make us all very vulnerable. Technology hasn't done away with the more severe widespread droughts for which stockpiling will prove inadequate. In good times like today, populations expand to fill the space available within only a few generations. And it isn't just physical space: as farmers manage to feed more people from each acre of land, population growth fills in that "excess capacity." Let the productivity slip for any reason, such as a worldwide drought, and people starve if their society hasn't devised good backup food sources. Birth control, by keeping the population levels to what can still be supported in the least productive years, would be a humane alternative to the boom-and-bust cycle.

Can we think our way out of this, produce yet another technofix? Birth control might do the job if there were a long transition into a drier climate, slowly downsizing the human population to track the resources. Though I doubt it, we might actually bypass the usual downsizing via the Four Horsemen of the Apocalypse, might even bypass what African history teaches us about the last thousand years of half-century-long droughts in otherwise good times. Hothouse agriculture might, were it widespread enough, even feed a larger population than today's. Given enough time to make a gradual transition, all sorts of technofix might be possible, much as in the story that some thinkers advance – the notion that the Earth might actually be able to support twice as many people as now.

Alas, they assume some form of stability that doesn't exist, as they would know if they bothered to learn the paleoclimate story. The transition-time factor cannot be left out of the equation. We are very vulnerable to sudden decade-scale change. When it happens, the struggle is not pretty.

THE SUN SETS RATHER QUICKLY in the tropics. It's dark already, with the distractions of the southern skies beginning to appear. This isn't at all the slow transition that I'm accustomed to, living at a latitude where twilight is prolonged. Sunset in the Serengeti reminds me of the Eurostar train popping into Paris, without the slow transition from farmland to city via protracted suburbia.

I hope that's not a metaphor for our future climate, that it will arrive more suddenly than we expect. But the record of past climates certainly makes it likely. Until a dozen years ago, everyone lived in blissful ignorance about the severe climate flips that happen quickly, with just a few short years before hard times set in with a vengeance.

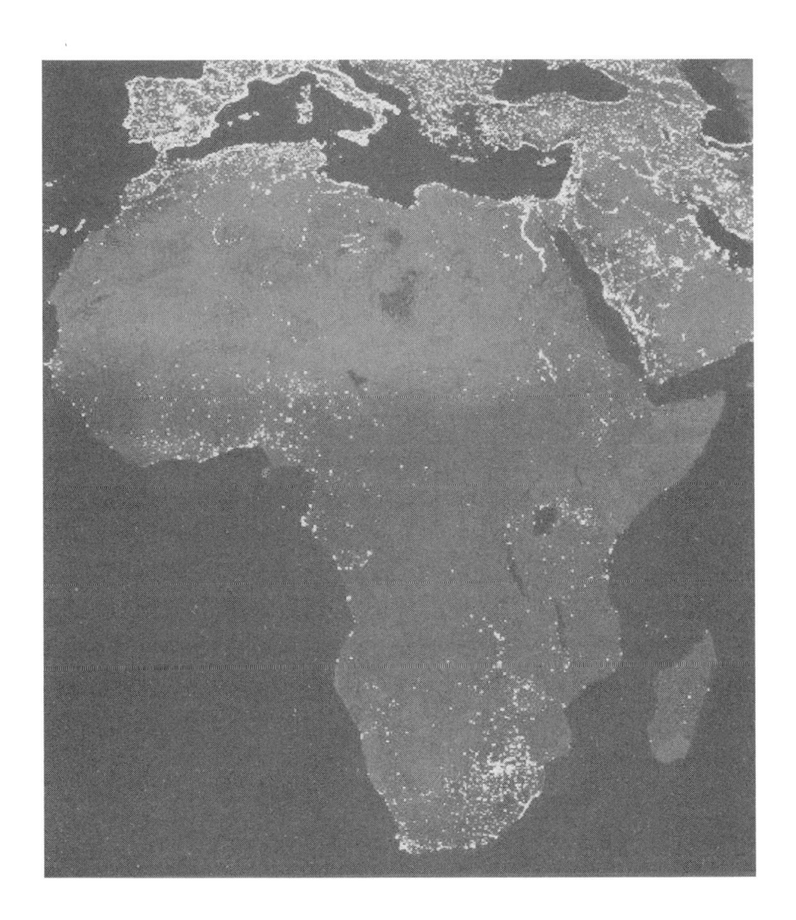

Africa at night

To:	Human Evolution E-Seminar
From:	William H. Calvin
Location:	29°N 22°E 9,100m ASL
	Libya by moonlight

Subject: **The last big step toward humans**

I just happened to look out of the airplane window in the middle of the night, and saw dozens of fires scattered across the moonlit landscape of Libya. No, they aren't the romantic campfires of the desert tribes, scenes from *Lawrence of Arabia*.

There is a lot of methane gas mixed in with crude oil, and they often burn it off rather than capturing it to export as natural gas. Just look at one of those posters of the night side of the Earth as seen from space and you'll soon learn where the oil fields are, from all the bright lights in thinly populated places like the shores of the Red Sea.

A methane torch isn't as satisfying as a nice campfire. There might, of course, be reasons why we find campfires so pleasing, particularly campfires in a cave. By dawn, we'll be over southern France, overflying a lot of famous caves.

TALK ABOUT SATISFYING VIEWS – I'm told that you can't drive around the river valleys in southwestern France without remarking on all the nice sites for rock shelters in the limestone walls of the valleys. The ledges, sheltered by overhanging rock, look out on water and grass – and were likely a superb place to spot grazing herds, back in the ice ages. I must remember to ask Gordon Orians if landscape aesthetics can distinguish between savanna with wide views and such limestone valleys with narrowed views, given that both could feature elevated viewpoints, scattered trees, grass, and edible grazing animals.

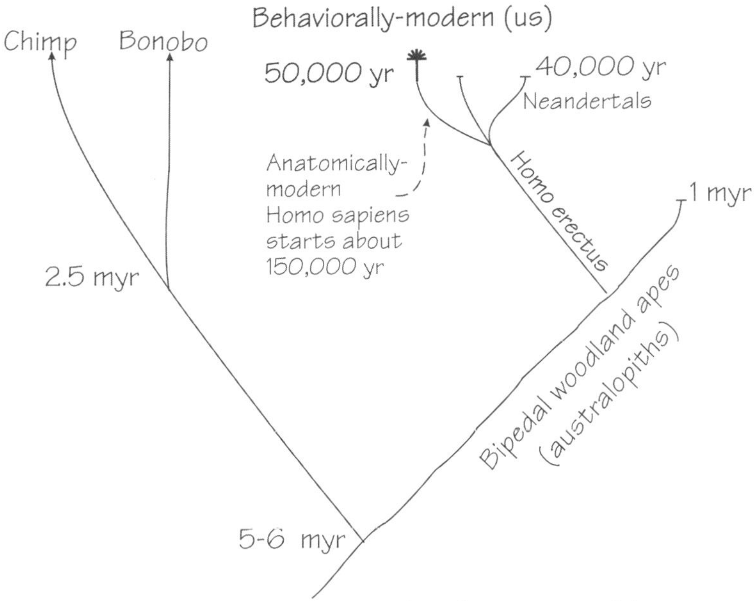

The rock art thereabouts is what draws most visitors to the area, not paleohunting tales or the Cro-Magnon fossil site. And it focuses us on another important development, perhaps the last big one in human biological evolution – or perhaps the first really big episode of cultural evolution. The number of big steps is surely arbitrary, but I'd say that there have been at least four during the last five-to-seven million years, the time since we last shared a common ancestor with the chimp and bonobo.

The first was attaining upright posture, what I discussed at Sterkfontein cave in South Africa, featuring the australopithecine skeletons.

Then comes adapting to the limited amounts of savanna, perhaps 2-3 million years ago. We might have had a number of changes in prolonging childhood and in social behaviors, including such things as pair-bonding.

Third is toolmaking (and likely some improvements in other plan-ahead behaviors, maybe even protolanguage). Toolmaking got off the ground by about 2.6 million years ago, and that's when the brain started to enlarge beyond the size of the great apes. Apparently it was Benjamin Franklin who coined that term "Man the Toolmaker" two centuries ago, but now it looks as if toolmaking complexity was not a long steady rise. It is widely suspected that tools were mostly relevant because of the addition of a lot more meat to the diet at this time. *Homo* changed in many ways: a jump in brain size of almost half back before 2 million years ago, a much less heavy body (australopithecines had apelike body mass for their height), and with males closer in size to females. Growth probably slowed back then, with a prolonged childhood.

Finally came what anthropologists call "behaviorally modern" humans. The beginnings of this, more than 50,000 years ago, were found west of Lake Naivasha (page 150). Since then, tools became much finer, fashioned from delicately struck stone blades, and fishing implements appear. Sewing needles are seen by the coldest part of the ice age. Bone and antler came to be used for sharp tools, and were even decorated with carvings. Cave art appears, as do beads and pendants. They probably thought a lot like we do, even though some still had brow ridges and bigger teeth.

My father used to say that, through culture, humans effectively domesticated themselves. As we know, domestication - of plants and animals - leads to rapid evolutionary change.
– RICHARD E. LEAKEY, 1992

Earlier than that, whatever they were thinking didn't often show up as durable art or as more sophisticated tools (except, notably, in Africa). As Ian Tattersall said in *Becoming Human*, the last 50,000 years "stands in dramatic contrast to the relative monotony of human evolution throughout the five million years that preceded it. For prior to the Cro-Magnons, innovation was . . . sporadic at best."

Now if only we could measure it in the archaeology, a better set of criteria for modern human behavior would include abstract thinking, planning depth, innovation, and symbolic behavior. I formulate it a different way, as the suite of structured thought called higher intellectual function, but they largely overlap.

THE BRITISH ANTHROPOLOGIST KENNETH OAKLEY suggested in 1951 that this efflorescence might have also been when fully modern language appeared on the scene, and that's still the leading idea, as I mentioned back at the Maasai Mara. While there were likely earlier forms of communication, "Only language," to quote the linguist Derek Bickerton [he means, of course, syntax - not just words], "could have broken through the prison of immediate experience in which every other creature is locked, releasing us into infinite freedoms of space and time." (Though, some psychotherapists note, there remain a few modern people who still cannot break through the prison of immediate experience.)

I suspect that this cultural efflorescence followed the step up to the modern suite of structured mental abilities that I call the higher intellectual functions, of which syntax is the prime example. It's not clear how long the higher intellectual functions have been around, which is why people talk of anatomically-modern *Homo sapiens* and then behaviorally-modern *Homo sapiens*. Judging from bones and DNA, our species is now thought to go back perhaps 150,000-200,000 years, and Africa's Middle Stone Age starts at about 250,000 years ago.

The strong signal from the behavioral record, then, is that our acquisition of the human capacity was a recent, and emergent, happening. Much as paleoanthropologists like to think of our evolution as a linear affair, a gradual progress from primitiveness to perfection, this received wisdom is clearly in error. We are not simply the inevitable result of a remorseless process of fine-tuning over the eons, any more than we are the summit of creation.
– IAN TATTERSALL, 1998

The most spectacular evidence for lively mental abilities comes from cave paintings. The Chauvet cave in France is dated to 33,000 to 38,000 years ago and an Italian cave may be slightly older yet. The cave paintings speak, as Richard Leakey says, "of a mental world we readily recognize as our own." It, like the scattered use of diagrams and bone, could go back earlier. Because Chauvet has fully representational paintings, not primitive stuff, art may go back much earlier in other undiscovered places. As the archaeologists like to say, "Absence of evidence is not evidence of absence" and it is surely an appropriate caution where ephemeral art forms are concerned.

The dichotomy (if there is one; it might disappear as the archaeology gets better) between anatomically and behaviorally modern *Homo sapiens* is opposite of the situation we usually face. In general, it is best not to identify innovations with the species that obviously practiced them. Big innovations (say, upright posture) are likely invented earlier by the predecessor species. Behavior invents, bones follow (a less elegant way of saying *form follows function*). In this view, the bony transitions are merely solidifying progress already achieved by behavioral flexibility and learning abilities, likely including cultural transmission. But in the case that the cave art illustrates, the bony changes could be 100,000 years earlier. This is behavioral innovation without (as yet) a corresponding change in body proportions and other things that paleoanthropologists focus upon.

THE LAST BIG STEP UP may turn out to have something to do with augmenting "primary process" in our mental lives. The notion is that it got an addition called "secondary process." The distinction was one of Freud's enduring insights, though much modified by now.

Primary process connotations include simple perception and a sense of timelessness. You also easily conflate ideas, and may not be able to keep track of the circumstances where you learned something (what the memory experts call "source monitoring"). Primary process stuff is illogical and not particularly symbolic. You consequently engage in a lot of displacement behaviors like kicking the dog, when frustrated by something else. While one occasionally sees an adult who largely functions in primary process, it's more characteristic of the young child who hasn't yet learned the differences in point of view between itself and its mother. There is a lot of automatic, immediate evaluation of people, objects, and events in one's life, occurring without intention or awareness, driven by the immediacy of here-and-now needs and having strong effects on one's decisions and behavior.

Secondary process is layered atop primary process, and is much more symbolic. Language is used to relate experiences to someone with, commonly, a different viewpoint. Secondary stuff involves symbolic representations of an experience, not merely the experience itself. It is capable of being logical, even occasionally achieves it. You can have extensive shared reference, the intersubjectivity which makes institutions possible. Piaget's developmental stages show you another way of looking at what might have also been evolutionary stages.

NOW LET ME TACKLE what happens in the Upper Paleolithic (in Europe; in Africa, the Middle and Later Stone Ages), and why. The "50,000 year problem" is that some modern mental abilities become apparent in art and toolmaking between 50,000 and 35,000 years ago, with the stone-tool improvements visible much earlier in Africa's Middle Stone Age. While there's no big change in gross anatomy, that doesn't rule out some brain and behavioral change caused by a changed mix of genes - such as improvements in language and planning abilities.

Before then, did anatomically-modern humans just talk silently to themselves, like the animals in Gary Larson cartoons? Most people assume they did. They also believe that our pet cats and dogs think, and certainly apes. So let me tell you what I always say to such statements about the great apes: If they could talk to themselves with the complexity we seem to assume, then they could think complex thoughts. And if they could think complex thoughts, they'd be able to plan ahead and do other things clearly advantageous to themselves. We'd see the evidence for complex thought in their behavior, even if they didn't talk about it.

But after you've learned enough animal behavior to also know what they don't do, you begin to wonder. If chimps had complex thought, for example, they'd be the terror of Africa. Instead of their ganglike hit-and-run raids on an isolated neighbor, chimps would make war on whole groups of neighbors using stockpiling of supplies and staged, coordinated attacks. We do not see that. No one sees much evidence of logical planning in the chimps, not the kind of planning where two or three novel stages need to be worked out in advance of acting. No such evidence in chimps, and certainly not in Gary Larson's talking snakes.

Since the great apes don't much plan for tomorrow, I'm willing for the moment to consider that complexity of thought may not be present in them. And maybe this same viewpoint ought to be applied to our ancestors, too, at least considering the possibility that complex thought is not much older than 50,000 years. Before then, *Homo sapiens* would have been unapelike in many respects. They would have had the long childhood with lots of culture to learn; they might well have had protolanguage, with its vocabulary and short sentences. Their feelings might have been much like ours.

But without structured thought, with its tendency to spin scenarios about the future and concoct explanations for the past, they might have lived largely in the present. Their logical abilities might have been limited to what we see in apes, and their sentences might have rarely been chained together for a larger effect. They might have lacked that horizon-expanding addiction to stories that we have, or our tendency to spend years trying to fulfill an idea.

MAYBE NEW GENES APPEARED ON THE SCENE 50,000 years ago. Maybe not, too – we're always happy to credit some new gene, but it's even more likely to be the loss of an allele than the addition of one. It's the committee of genes that changes, the committee that regulates early brain development trajectories and staging, from whence comes the behavioral propensities.

The Y chromosome data suggests a major branching about 50,000 years ago, as does the mitochrondrial DNA dating, raising the possibility of some late genetic change affecting higher intellectual function.

One line of Y chromosome evidence suggests that there was a migration from central Asia into Europe between 50,000 and 35,000 years ago. It may have brought with it what the archaeologists call the Aurignacian, an advanced culture known for its sophisticated rock-art paintings and finely crafted tools of antler, bone, and ivory. Maternal mtDNA allows some similar inferences about migrations. While neither mtDNA nor the Y chromosome may have carried the behavioral genes that made the difference, they do suggest when some founder effects might have marked the transition from anatomically-modern to behaviorally-modern *Homo sapiens.*

It's still "early days" on judging bottleneck dates, with different genes giving different results. One is reminded of the most famous "young date" of all, when Lord Kelvin calculated that the earth couldn't possibly be as old as Darwin and the geologists said (his calculations were fine but based on the assumption that the earth wasn't heating itself – no one yet knew about radioactive decay and the energy it releases, or how poorly the deep rocks conduct heat).

We too are likely missing something that will change our calculations. That's just science. But let me take that 50,000 year date seriously for a moment, as there is an interesting window of opportunity about then for moving large numbers of people out of sub-Saharan Africa up north to the shores of the Mediterranean.

THE SAHARA IS ORDINARILY A GREAT BARRIER to any journey that requires food and water to be found along the way. Especially during the cooler-and-drier parts of the ice ages, essentially no rain was thought to fall there. As I mentioned earlier (at page 63) when flying over the Sahara and Sahel, there are pluvial periods when the monsoon rains extended many degrees farther north into the Sahara and grass and bushes grew everywhere. Various factors contribute to the extent, but the most obvious is that the tilt of the earth's axis varies about 2.5° over a cycle of about 41,000 years, and 9,500 years ago it reached its peak of 24.6°.

The prior peak of this tilt cycle would have been about 50,000 years ago (there's a plot on page 207). Of course, the cool and dry climate of Africa during the middle of an ice age might have kept maximal tilt from effectively producing a prior Pluvial at 50,000 years. But there were a half-dozen periods about then, lasting for centuries, when the worldwide climate kicked out of cool-and-dry into warm-and-wet within several decades time. They could have interacted with the maximal tilt's more penetrating monsoons to produce brief wet mini-pluvials in the Sahara (and no, my pulsed pluvials couldn't have happened back at 91,000 years ago because there were no flips back then - the plot is on page 223).

In the Sahel cycle, cattle can be grazed farther north for some years (as in the 1950s) but soon a drought (as in the early 1970s) pushes human populations back south - not north or east, because the Sahara itself remains such a resource barrier. During a period where even the Sahara has grasslands and lakes, one can imagine a window of opportunity for a sub-Saharan hunting population spreading all around the Sahara to exploit the boom-time population of grazing animals, only to be pushed out when the next flip abruptly ended the Sahara's temporary resource boom.

The Sahara can thus be seen as a pump, drawing populations in from the Sahel and then abruptly pushing them out a few centuries later - and not merely back into the Sahel, but with some of them fleeing the profound drought along the Mediterranean shores. A half-dozen quick strokes on this pump may have occurred starting about 56,000 years ago, with a more prolonged interstadial-length warming of a few thousand years centered around 48,000 years ago. A series of pulsed pluvials is far more effective than a single 9,000-year-long pluvial in spreading hunters and herders from Africa into the grasslands of Asia.

IF OUR MODERN LIFE OF THE MIND is less than 50,000 years old, however, it does raise some interesting questions. The efficiency notions surrounding evolution often lead to generalities about how evolution produces "well-tested" parts. (This, despite all of the evidence from medicine about how poorly "designed" a lot of things are, such as the all-important female reproductive tract with its leaky seal between ovary and Fallopian tube that allows ectopic pregnancies and endometriosis.)

But when biological innovations are still young, then all bets are surely off because evolution simply hasn't had much time to get the bugs out. We don't expect reading to work well because it has only been around for 5,000 years and then only in a few percent of the population (until recent centuries). Five thousand years is only fifty centuries and, at four generations per century, that's a mere 200 generations.

Human beings are animals. They are sometimes monsters, sometimes magnificent, but always animals. They may prefer to think of themselves as fallen angels, but in reality they are risen apes.
– DESMOND MORRIS, 1967

Well, 50,000 years is only 2,000 generations and that sure isn't much time, not when comparing to that hundred-fold greater span of 5-6 million years since the common ancestor with the chimps and bonobos. While structured thinking may be one of the aspects of human uniqueness, I suspect that evolution hasn't yet tested it very well - that it is chunky and perhaps dangerous, both to ourselves and the other inhabitants of our planet. Perhaps evolution didn't have time to get the bugs out before distributing it worldwide (but let me save that topic for another E-seminar, when I visit the cave art we're flying over about now).

It also isn't clear that we're good enough yet to craft a science and technology that will head off a big climate catastrophe. We are, at least, able to perceive the problem now. And we have the computers to help us evaluate the do-something scenarios.

SOMETIME NOT LONG after the efflorescence, humans floated to Greater Australia (at low sea level, New Guinea is connected) and, somewhat later, maybe even to the Americas. There are sites in South America dating back to 20,000 to 30,000 years ago, and more recent ones in eastern North America. Given that getting across the oceans or the Arctic required a lot of planning ahead, the journey may have been made possible by the improved suite of higher intellectual functions.

The really definitive spread into the Americas, the one that may have overridden any surviving earlier settlements, occurred as the last ice age was coming to a close. It is thought that some hunters arrived in Alaska about 15,000 years ago during that abrupt warming episode known as the Bølling-Allerød, where the Northern Hemisphere suddenly warmed up to almost modern levels (and in only a few decades, too). Since it took much longer for sea level rise to follow, it would have been a good time to pursue those elephants we call mammoths across the exposed Bering Strait, and discover an even bigger supply.

But the continental ice sheets probably kept them from spreading south through Canada until an ice-free corridor opened up 13,000 years ago from Alaska to Montana. There were two major ice sheets, the one over the Rockies and the one centered on Hudson Bay which had spread west into the foothills of the Rockies. As things warmed up, the latter ice sheet retreated and some grass started to grow in the gap. This provided food for various Alaskan species, allowing them to gradually migrate southward in the corridor along the eastern foothills of the Rockies.

The Alaskan brown bear, the grizzly, made it down south about then. So did some big game hunters, about 13,000 years ago. Though they weren't the first, they discovered an even more plentiful supply of mammoths, which were then living all over the Americas. It took less than a thousand years for those hunters to spread all through North and South America. That may sound fast, but it's only eight miles per year - and a hunter is likely to wander as far in a single day. A thousand years was also about the time that it took to kill off all the mammoths (and other megafauna) in the Americas, leaving the hunters with an urgent need to diversify their diet. Note that an initial population of even a hundred people could have grown to 10 million in that same thousand years, were there a growth rate of only one percent per year (Kenya's present population growth is three times as fast, for a doubling time of only 23 years).

Exactly who these early Americans were is a subject of debate, as their skull shapes are sometimes more like those of people from Europe and Africa than they are like those of the northern Asians living closer to the Bering Strait. The present-day Native Americans are mostly from later waves of immigration from Asia, with the Arctic-adapted Inuit ("Eskimo") of Canada and Greenland being the most recent.

And that's where I'm heading now, to take a look at what happens in the oceans near Greenland's eastern and southwestern coastlines, the processes that make worldwide climate flip so abruptly from warm-and-wet into cool-dry-dusty-windy.

To: Human Evolution E-Seminar
From: William H. Calvin
Location: 52.309°N 4.767°E -2m ASL
Layover Limbo (again)

Subject: **The Little Ice Age and its witch hunts**

Our civilizations began building immediately after the last great continental ice sheets melted about 10,000 years ago, thanks to agricultural success. Our present warm period has been relatively mellow for the last 8,000 years, with fewer fluctuations than any other similar period in the climate record. The "Little Ice Age," a cooling of less than one degree that lasted from the early Renaissance to the middle of the nineteenth century, was small stuff in comparison to the ice ages or the abrupt jumps.

But the Little Ice Age has a human scale. You can recognize it by extrapolations from bad storms and droughts in 20th century newsreels and television. "The Little Ice Age," the archaeologist Brian Fagan writes, "reminds us that climate change is inevitable, unpredictable, and sometimes vicious." So it is worth a few pages on the Little Ice Age and contemporary crop failures, just to give you an emotional base, one from which you might conceivably manage to further extrapolate to abrupt climate flips lasting centuries with tenfold greater temperature excursions and far more widespread disruption of ecosystems.

THE LITTLE ICE AGE as a term is a little vague. Its first recorded use in 1939 was a very informal analogy, not a serious attempt at naming, and it covered the last 4,000 years! The term probably caught on as a description of 1300-1850 because our images of those times came from paintings of ice skaters on the rivers of Europe and tales of people walking across the frozen

Baltic between Denmark and Sweden. Books on the subject feature the detailed wintertime scenes painted by Peter Breughel the Elder during the first great winters of the Little Ice Age about 1565. An exaggerated winter seems to be what we carry away from the subject, what remains after we have forgotten the details. Even historians are unhelpful. You can read an excellent history of the last five centuries, such as Jacques Barzun's *From Dawn to Decadence,* without finding the Little Ice Age mentioned even once.

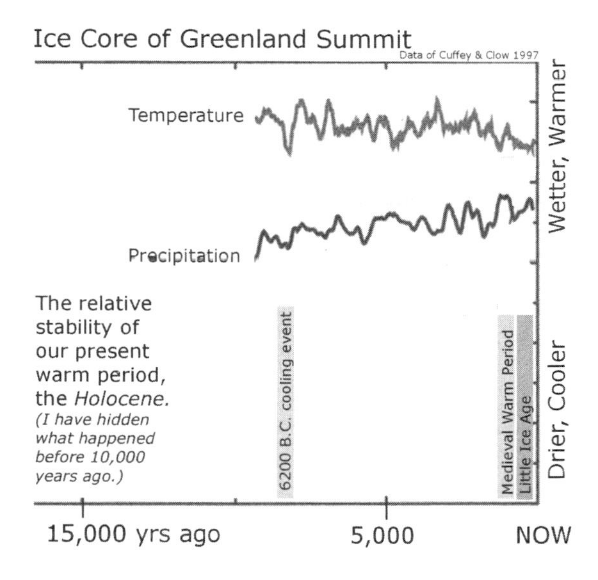

And the average temperature changes of the Little Ice Age seem small - what's a degree or so, anyway? None of us are used to dealing in yearly average temperatures, and the annual mean temperature hides the combination of a cold winter and a hot summer. Because the means are never reported, we have no sense of how much a warm year differs from a cooler year. We have the same problem in judging how much difference several degrees of global warming will make to our lives.

By focusing on temperature per se, we conveniently ignore the more important aspects of climate change: floods, droughts, high winds, dust storms, and unseasonable weather that ruins harvests and sets up famines. Unsettled extremes are also what the Little Ice Age was all about. They are what cause people to starve, not what the thermometer reads.

The most obvious causes of droughts are changes in the winds, what happens with almost any climate change scenario. Afternoon heating of inland regions causes air to rise, attracting in moist air from offshore which, upon warming and rising itself, dumps its moisture. Such monsoon winds often fail for years. Millions of people in India died in the monsoon failures of the late 19th century, and relief efforts in earlier such famines were hampered by the lack of transportation infrastructure.

In the 1930s, there were particularly strong westerly winds in the U.S. This extended the rain shadow of the Rocky Mountains farther eastward and reduced the monsoons from the Gulf of Mexico, resulting in the Dust Bowl conditions.

Lack of winds can also block rains: a persistent high pressure system, not an uncommon occurrence, blocked weather systems that might have brought rain to the American Midwest and East in the summer of 1988. Searing heat caused at least half of the grain crops to be lost in the northern Great Plains states. Long stretches of the Mississippi River became so shallow that barges were stranded on mud banks for weeks. Huge forest fires burned major sections of Yellowstone National Park that year.

When this much damage can be done in a single year, just remember that serious droughts endure for decades. When people are told to stop watering their lawns because of a water shortage, they escalate (in the manner of sports hyperbole) to use the same word, *drought,* as is used for far more serious conditions, on a far vaster scale and lasting many years - such as the 1930s Dust Bowl or those three Little Ice Age droughts amidst good times in East Africa, lasting 30, 65, and 80 years.

Besides lack of rain, famines can occur from other climatic causes, thanks to our reliance on agriculture to support a thousand times greater population than before agriculture. Too much rain at the wrong time can flatten wheat fields, one of the reasons why the Irish shifted from a reliance on grains to less chancy potatoes. Potatoes were a huge success and the Irish population grew. But the over reliance on potatoes left Ireland vulnerable to a potato blight that crossed the Atlantic Ocean in the 1840s, causing (together with the ineptitude and unfettered trade ideology of the British government) the deaths of over a million Irish. Far greater famines occurred in the twentieth century exacerbated by politics, such as the Volga famine of 1921, the Ukrainian famine of 1932-33, and China's Great Leap Forward with famine deaths calculated at between 36 and 50 million from 1959 to 1961.

In addition to the miseries from famine and its associated diseases, people created their own miseries, driven by a frantic urge to "do something." Scapegoating surged as people blamed other people for their misfortunes. Americans tend to look with suspicion on the Puritans because of the Salem witch trials without realizing that this type of scapegoating was endemic earlier all across Europe and not peculiar to the Puritans. It coincided with the demoralization produced by the worst years of European climate change.

Fagan notes that "As scientists began to seek natural explanations for climatic phenomena, witchcraft receded slowly into the background." While I hope it stays there, remember that most people still believe in horoscopes - and that most Americans have only rudimentary science literacy. The rational aspects of civilization are a thin veneer and scapegoating, perhaps escalating to genocide, could still happen in future climatic crises.

Famine followed famine bringing epidemics in their train; bread riots and general disorder brought fear and distrust. Witchcraft accusations soared, as people accused their neighbors of fabricating bad weather.... Sixty-three women were burned to death as witches in the small town of Wisensteig in Germany in 1563 at a time of intense debate over the authority of God over the weather. . . . Between 1580 and 1620, more than 1,000 people were burned to death for witchcraft in the Bern region alone. Witchcraft accusations reached a height in England and France in the severe weather years of 1587 and 1588. Almost invariably, a frenzy of prosecutions coincided with the coldest and most difficult years of the Little Ice Age, when people demanded the eradication of the witches they held responsible for their misfortunes.

–BRIAN FAGAN, 2000

Nothing is more terrible than to see ignorance in action.

– JOHANN WOLFGANG VON GOETHE

Science is an attempt, largely successful, to understand the world, to get a grip on things, to get hold of ourselves, to steer a safe course. Microbiology and meteorology now explain what only a few centuries ago was considered sufficient cause to burn women to death.

– CARL SAGAN, 1996

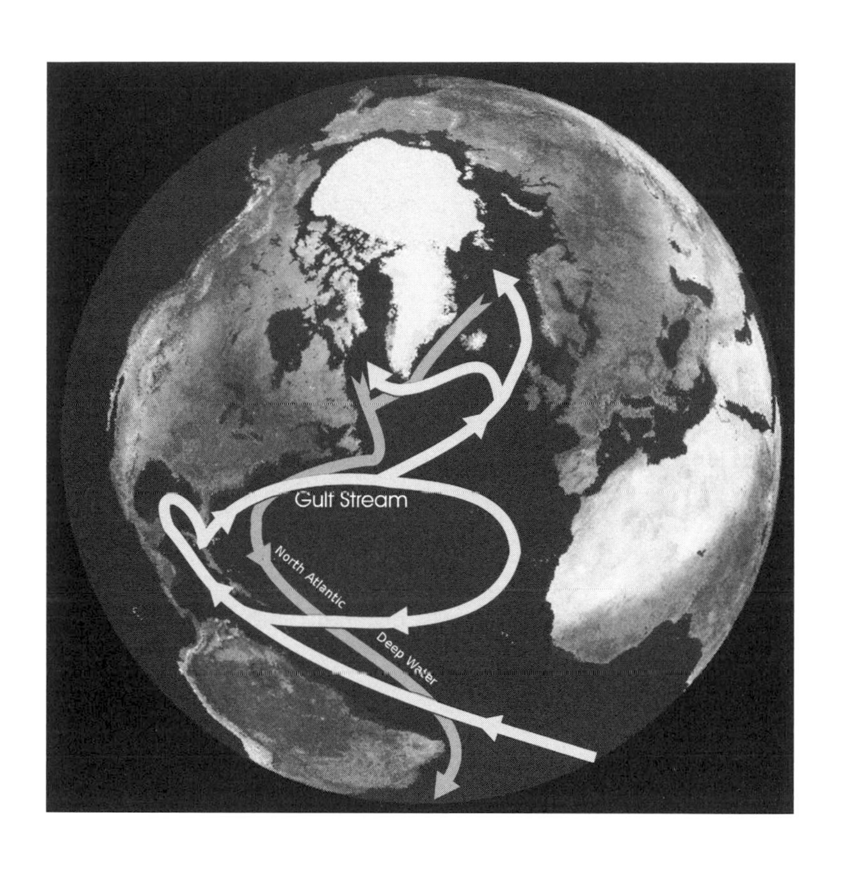
Gulf Stream
North Atlantic
Deep Water

To: Human Evolution E-Seminar
From: William H. Calvin
Location: 55.69222°N 12.55847°E 3m ASL
Copenhagen's ice cores

Subject: **Slow ice ages and abrupt whiplashes**

You wouldn't think that the direct route from Nairobi to Seattle goes over Greenland. But just try stretching a string over a globe's surface if you don't believe me. The shortest string goes right over the Sudanese and Libyan deserts, then over Copenhagen, northern Iceland, and Greenland.

I particularly like the view from the great circle route connecting Copenhagen to Seattle, so I have managed to get a "geologist's seat" (window seat, in front of the wing, on the shady side of the plane, take your binoculars). One flight in three, the clouds part and make your efforts worthwhile.

From Copenhagen, the plane goes over Norway and then the Greenland-Iceland-Norwegian Sea (sometimes called the "Nordic Seas" but I'll just call it all the Greenland Sea). Next it passes over a mountainous minor continent of rock and ice paradoxically called Greenland (a name devised by Eirik the Red to promote immigration, an early instance of deceptive advertising). Then we go over the great stores of methane frozen into the Canadian tundra (and apt to be released into the atmosphere by greenhouse warming, making things even warmer still). Those sites are all key players in climate flip-flop scenarios. They are intimately connected to our African past and our worldwide future.

WE'VE TALKED A LOT about abrupt climate jumps without mentioning very much about how they happen. The big scientific questions often have three major facets: what, how, and why. And before I can get much further into the how and why of things, I'd better flesh out the rest of the What's, the slow background changes which made the abrupt coolings such a surprise to the scientific world in the 1990s.

It seems strange to realize that even in the middle of the 19th century, hardly anybody knew about the ice ages. Even scientists still talked in terms of a biblical deluge. But as early as 1787 in Scotland, scientists realized that giant boulders were carried long distances and deposited in the midst of a landscape in which their kind of rock was utterly foreign. Even peasants in Switzerland subscribed to such an explanation. Charles Darwin, in his *Voyage of the Beagle*, wrote in 1839 of boulders carried along in icebergs. Louis Agassiz spent a lot of time looking at alpine landscapes. Those Swiss valleys looked as if they had been scoured, over and over. From his position as president of a Swiss scientific society in 1837, Agassiz elaborated the earlier notions and proposed that massive ice sheets had pushed their way around Europe, sometime in the not-so-distant past. Darwin provided further geological details on the subject in 1842.

The landscape evolved. Not only did the Alps grow long glaciers but giant ice sheets sat atop much of Northern Europe, flattening the Baltic landscape. Some piles of ice didn't even have mountains centered beneath them; they were like sand dunes, obstacles that grew as the laden winds ran into them.

We eventually learned that the warm periods in the ice ages last only about 10,000 years and that it's a long time until our next one is "due" – about 100,000 years. We learned that 32 percent of the land was covered by ice sheets in the last ice age, but that a warm period like today's reduces the ice to only 10 percent (mostly Greenland and Antarctica, but you can also see some serious ice sheets in Norway, Iceland, Alaska, Tibet, New Zealand, and the Andes).

How many ice ages have there been? Dozens. Up until about 800,000 years ago, the major meltbacks occurred about every 41,000 years. More recently, the 100,000 year rhythm has been the more obvious one.

THE SECOND GREAT INSIGHT concerned how sunlight might vary to help melt off the massive amounts of ice. These slow changes in ice mass are mostly caused by even slower changes in the summer sunshine at latitudes like Scandinavia.

Elaborating on the 1842 suggestion of the French mathematician Joseph A. Adhémar, Milutin Milankovitch proposed during World War I that the sunlight reaching the higher latitudes controlled the ice ages (he did his laborious calculations as a prisoner of war), and that slow changes in the earth's orbit were important.

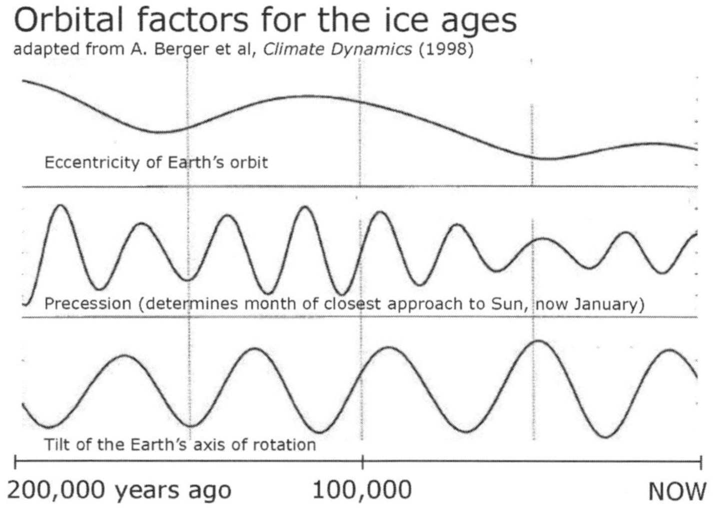

First of all, the tilt of the earth's axis varies over a 41,000 year cycle. Our axial tilt peaked 9,500 years ago at 24.6° and is now at 23.4° and heading toward a turnaround at about 22.1°. Furthermore, our closest approach to the sun in our elliptical orbit is currently in the first week of January (we're about 3 percent nearer, and get about 10 percent more heat, than in July) - but in another 12,000 years or so, the closest approach date will have drifted around to summer again, the precession helped by the gravitational pull of the other planets causing our less-than-spherical planet to slowly wobble. Our orbit is also more circular at some times (repeating at about 100,000 and 400,000 years), making the month-of-closest-approach factor periodically less important.

Tilt and precession account for a lot of the slow back and forth of the glaciers in between the major meltoffs, which occur every 100,000 years or so. The best setup for melting ice is when there are colder winters but hotter summers in the Northern Hemisphere - that is, after all, where most of the meltable ice is, thanks to the general lack of land between 40°S (southernmost Australia) and Antarctica. (Yes, I know, there are the southern Andes and some nice glaciers in southern New Zealand, but that's not much when compared to the land mass of Eurasia and North America north of Madrid and Kansas City.)

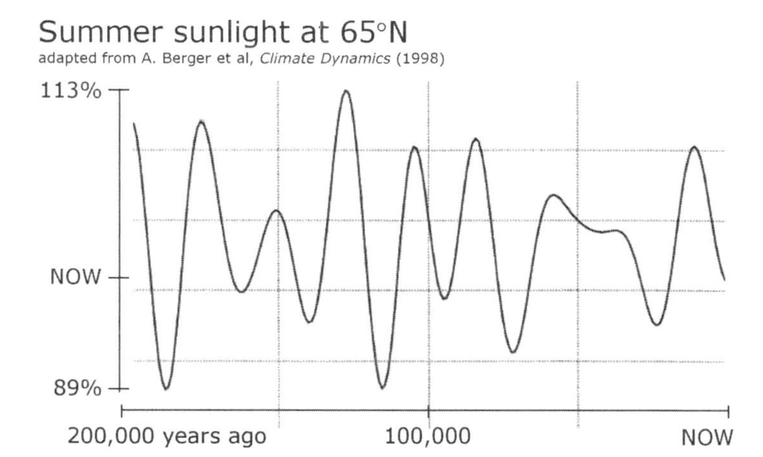

"Exaggerated seasonality" occurs when the closest-approach bonus is in summer *and* there's also another bonus from near-maximum tilt. In such cases, there is as much summer sunlight every day at 65° (say, in northern Iceland or Fairbanks, Alaska) as there is presently at 49° (say, in Paris or Victoria). Despite the accompanying colder winters, getting melting going during those long hot summers is how we got rid of the ice sheets at high northern latitudes. At the opposite extreme, 65° gets about what Thule (78°N) gets today. These rhythms, and the way they reinforce one another occasionally, also explain a lot of the back-and-forth between one major meltoff and the next, the so-called interstadials.

BUT IT WAS LONG SUSPECTED that "orbital factors" weren't the whole story, as the southern hemisphere ice sheets melted back at the same time as the northern ones. Why should ice sheets in the Andes and New Zealand melt when the closest-approach bonus was in their wintertime? They should have been out of phase with the northern one, but they were in phase, synchronized by something.

The other thing that made scientists suspect that it wasn't so simple was a wildflower, white with a yellow center, of the rose family called *Dryas octopetala.* It is cold-adapted and is found on the tundra - not among shrubs and pine trees. Well, a century ago, *Dryas* pollen turned up in cores of a lake bed in Denmark above a layer of trees, at a depth now dated back to about 12,000 years. A return to cold made no sense then, according the Milankovitch orbital view. That was when the two astronomical bonuses were combining to produce particularly hot summers in Denmark. Half of the ice sheets covering Europe and Canada had already melted, and all of the ones in Scotland. There should have been pine seeds in those cores, not the pollen of a tundra flower.

This return to ice age temperatures, called the Younger Dryas, lasted more than a millennium, reestablishing glaciers in Scotland once again. Then, things suddenly warmed back up. Nothing in the orbits could explain such things.

IF I HAD VISITED COPENHAGEN researchers in the late 1970s, no one would have agreed on when the ice ages began. The textbooks, noting the age of old moraines plowed up by the ice sheet frontiers, would have said the ice ages went back about 800,000 years. No one yet knew about winter sea-ice reaching the British Isles 2.51 million years ago, nor about all those antelope speciations in Africa about the same time.

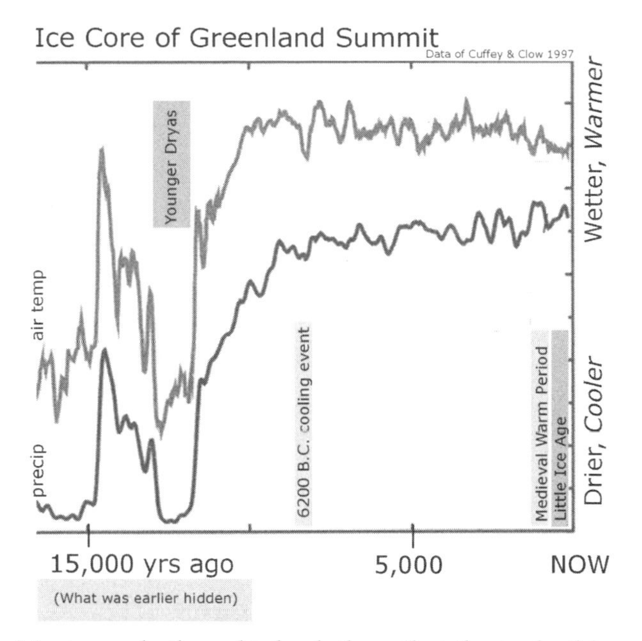

Most people thought, back then, that the ice buildup and the ice melting were largely due to those slow changes in various aspects of the earth's orbit around the sun. Since they don't suddenly jump, most people assumed climate would change gradually. The cores coming out of the ocean floor indeed showed slow changes, and so did the ice core from Antarctica - but few understood then about how special Antarctica is, how insulated from the rest of the world its weather can be (that ring of westerlies and its vertical curtain of rising air at 60°S) and how its low rates of snowfall limit time resolution. (Various parts of Antarctica are now known to give different results.)

But the pros knew about such anomalies as the Younger Dryas and they were anxious to get better data. Greenland had more snow each year, thanks to weather systems sweeping up from the Gulf Stream, and the Greenland cores were starting to show some puzzling results, as the Danish researcher Willi Dansgaard and colleagues reported in 1982. I first heard of the abruptness, per se, in 1984 when the Swiss geophysicist Hans Oeschger gave a talk in Seattle. The time calibration on one of his slides prompted me to ask him afterward, about just how quickly temperature had changed.

Oh, he said, the big drop took just a few years. The enormity of such a whiplash caused me to assume that we were having some language difficulties and so I persisted, asking, "Just a few *decades*?" No, no, he replied, merely a few *years*.

The American geochemist Wallace Broecker also heard Oeschger give a talk in 1984, and the quick flips gave Broecker his idea for different modes of circulation via failures of the salt conveyor (more in a minute). It was Broecker who coined the term "Dansgaard-Oeschger events" to describe the abrupt coolings and re-warmings. (The D-O's are what I've been calling the flip-flops or abrupt jumps).

Another important discovery was from the ocean-floor cores of the North Atlantic. Now and then, a rain of rock had fallen down to the abyss, dropped off the bottom of passing icebergs. The rocks mostly came from Hudson Bay and the "rain" was heaviest in the Labrador Sea. These "Heinrich events" are not completely understood. They do tend to occur in the coldest parts of the glacial cycles, when ice sheets extend out on to continental shelves. Perhaps they represent glacial advances that overrun terminal moraines and then freeze their rock into the bottom of the ice sheet. Later, when tides manage to float a terminal glacier and break it off at a weak "hinge," an iceberg is created which sets sail for the Bay of Biscay, held upright by its rocky ballast on the bottom. As it melts en route, so a trail of debris is deposited on the ocean floor – and fresh water is leaked all over the surface of the mid-Atlantic Ocean. That the events are episodic, lasting for several centuries, is attributed to the Hudson Bay ice mountain collapsing, perhaps from melting on the bottom in the manner of subglacial lakes in Antarctica.

The D-O flips are not the same thing as the Heinrich events, though the issue was confused for awhile. For example, a D-O cooling occurred in the midst of the last warm period at about 122,000 years ago, when there were no ice sheets in Canada to create icebergs. But in icy times, Heinrich events do tend to be shortly followed by an abrupt D-O warming.

SUMMER IS DRILLING SEASON and Greenland is where most of the Copenhagen researchers currently are. They try to extract long cylinders of ice, ones that will show annual layers of snow and ice. Greenland has ice that goes back to about 250,000 years ago, when it was warm for long enough that it melted most of the accumulation from prior ice ages.

Still, that quarter-million years seen in the ice cores contains the last two ice ages. And it includes the last warm period between 130,000 and 117,000 years ago, when there was a climate somewhat warmer than today's for much of the 13,000 year period. (For Europe, this warm period is called "The Eemian" - more generally, "marine isotope stage 5e.") Sea level was 3-5 meters higher than at present (just imagine shrinking southern Florida). For comparison, complete melting of the West Antarctic ice sheet (or, for that matter, the Greenland ice sheet) would today raise sea levels by 6–7 meters, which is several stories high. Melt all the ice sheets and it's more like 60 meters, covering up 18-story Florida condos, with the coastline somewhere up in Georgia.

The last ice age per se had several dozen D-O cycles, a flip one way or the other occurring on average about every 750 years. Our current warm-up, which started about 15,000 years ago, began abruptly in the Northern Hemisphere, with the temperature rising sharply while most of the ice was still present. Several thousand years later, temperature declined abruptly into the Younger Dryas. After the sudden rewarming at 11,550 years ago that ended the Younger Dryas, things gradually warmed into the period of modern sea level and agriculture.

There was a slow-and-gradual view of things in the good old days, back before we got the time resolution to see the abrupt stuff that our ancestors suffered through. These slow "orbital" causes of ice fluctuation have little to do with the rapid back-and-forth of a whiplash. Something else seems to be the more immediate cause of the chattering seen atop the slower orbital trends of ice.

Because some of the D-O cold-and-dry flips last for many centuries, there is a semi-slow way of viewing them, and indeed a group of European researchers are looking at the cluster of flips between 60,000-25,000 years ago for clues about the demise of the Neandertals, calculating the expected regional climate changes in Europe and correlating them with the dating of archaeological sites to see how the Neandertal populations were moving around in response (see their database web pages at *http://www.esc.cam.ac.uk/oistage3/*).

But my bust-then-boom aspect is keyed to the initial part of the abrupt changes, just the first century or two after a cooling-and-drying (when the drought-to-fire-to-grass-to-herds opportunity occurs) and the century after the abrupt rewarming that ends it (when grasslands extend into formerly arid areas, and again dramatically expand herd sizes). That first century after a flip should be when unusual opportunities present themselves for experienced hunters to expand their territory and populations. Archaeological time resolution cannot, at present, see such decade-to-century transitions, but that's likely where the action is.

Grounded ice sheets shown; they are generally connected by sea ice.

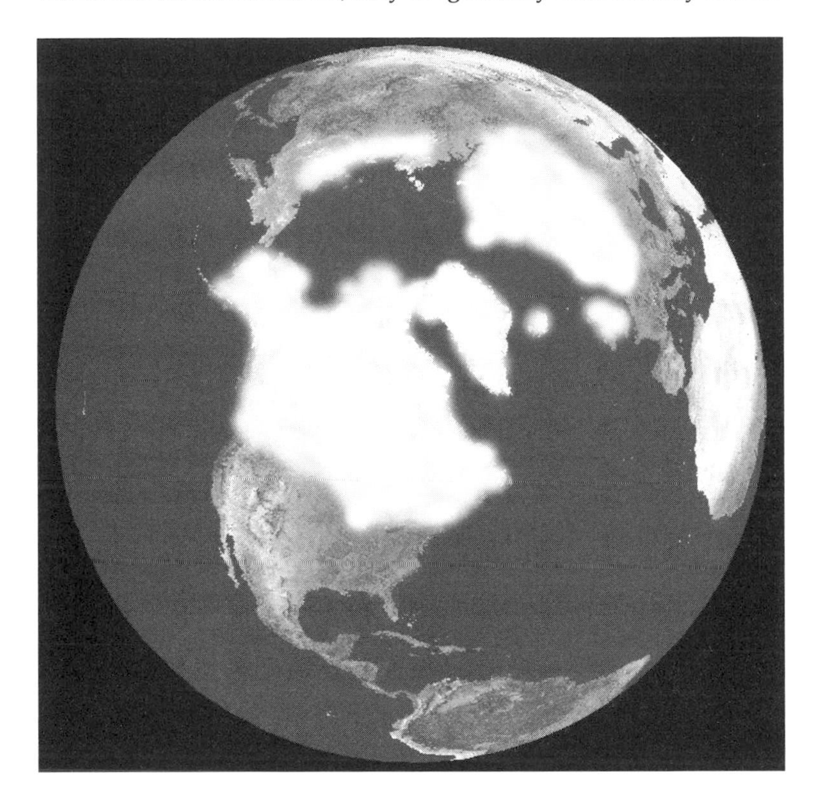

To:	Human Evolution E-Seminar
From:	William H. Calvin
Location:	55.620°N 12.656°E 2m ASL
	The plane where it's always noon

Subject: **How ice age climate got the shakes**

Archaeologists have always been interested in ancient climates, given how droughts tend to contribute to the demise of one civilization after another. There are many indicators of ancient climate, such as wind-blown soils piling up during periods of great wind storms. When they blow from Africa into the Atlantic Ocean, those that don't blanket the Cape Verde Islands sink to the ocean floor, where a core can be drilled to sample millions of years of Sahara dusty periods.

Coring old lake beds and examining the types of pollen trapped in sediment layers led to the discovery of the Younger Dryas early in the twentieth century. Pollen cores are still a primary means of seeing what regional climates were doing, even though they suffer from poorer resolution than ice cores (worms churn the sediment, obscuring records of all but the longest-lasting temperature changes). When the ice cores demonstrated the abrupt onset of the Younger Dryas, researchers wanted to know how widespread this event was. The U.S. Geological Survey took old lake-bed cores out of storage and re-examined them.

Ancient lakes near the Pacific coast of the United States, it turned out, show a shift to cold-weather plant species at roughly the time when the Younger Dryas was changing German pine forests into scrublands like those of modern Siberia. Subarctic ocean currents were reaching the southern California coastline, and Santa Barbara must have been as cold as Juneau is now. (But the regional record is poorly understood, and I know at least one reason why. These days, when one goes to hear a talk on ancient climates of North America, one is likely to learn that the speaker was forced into early retirement from the U.S. Geological Survey by budget cuts. Rather than a vigorous program of studying regional climatic change, we hear the shortsighted, preaching of cheaper government at any cost. These penny-wise-pound-foolish factions of the U.S. Congress even tried hard in 1995 to eliminate the U.S. Geological Survey altogether.

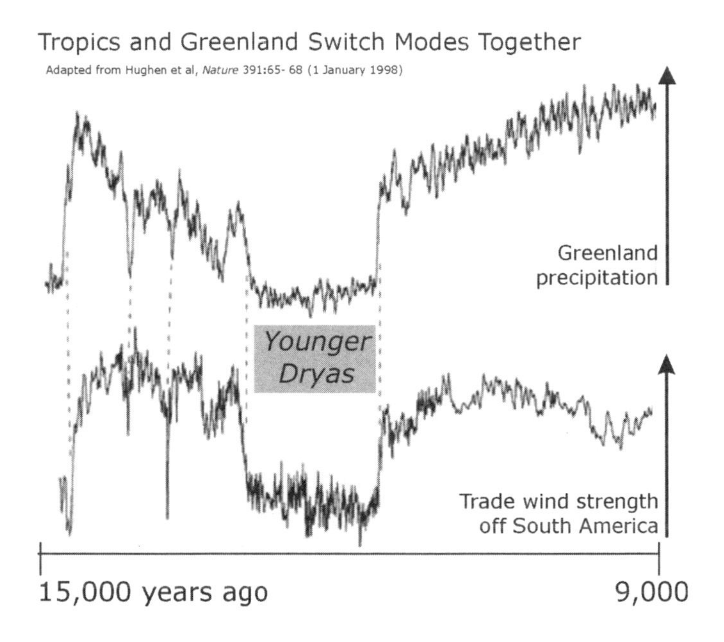

THE MAIN PROBLEM with most paleoclimate indicators is that time has gotten smeared out, so that events that last less than a thousand years simply disappear into the noise. On land, people tread paths into the floors of caves, constantly rearranging the deposits of previous centuries. At the bottoms of lakes, worms churn things. In the oceans, the flat fish burrow and stir.

Fluctuations, alas, can be very important, as in the story about the statistician who drowned in a lake whose average depth was only three feet. (And in the stock market, even a two-year moving average will hide significant swings, during which fortunes were made or lost.) For paleoclimate indicators, few have time resolution better than one millennium.

Great coolings or warmings could have happened and you'd never know it from such an averaged record (unless the event lasted more than a millennium, which several did). There are a few sites, such as the Santa Barbara Basin, with high accumulation rates (and anoxic bottoms to discourage the worms) that have resolutions that occasionally achieve fifty year discriminations. Ice has the best time resolution of all, especially at sites (such as southern Greenland) where a lot of snow falls each year. Ice cores can be made near the poles and at a few elevated low latitude locations as well.

Once the ice cores from Greenland were analyzed in the 1980s, experts such as Willi Dansgaard and Hans Oeschger made us aware that big warmings or coolings had occurred in less than a decade. Greenland does have some weather patterns that affect regional temperatures, but the big rapid transients recorded in Greenland are also seen in the tropics and in many places in the southern hemisphere, and at about the same time. If the prior ice ages turn out to be anything like this last one - and all suggestions so far point that way, with evidence for the Heinrich events already back to 0.8 and 1.1 million years - then our hominid ancestors suffered through hundreds of these dramatic events.

OUR PRESENT WARM PERIOD has been free of abrupt changes, at least since 6,000 B.C. An abrupt cooling got started 8,200 years ago, but it aborted within a century or so, gradually warming back up instead of stepping up. I'll mention the cause of this aborted cooling when we get a little closer to the scene of the crime. It probably corresponds to the drought seen about then in southeastern Europe and the eastern Mediterranean. Temperature changes since then have been gradual in comparison.

Indeed, we've had an unprecedented period of climate stability (nothing much worse than the Little Ice Age - though, of course, droughts of only a few decades duration were sufficient to ruin civilizations of the past). It may seem paradoxical to talk of the relative stability of climate in the last 8,000 years, and then mention the devastation caused by decades-long droughts and El Niños, but the latter are truly small stuff compared to the abrupt climate flips with their synchronous worldwide effects.

Why the relative stability for eight millennia? No one knows, yet. But we know it is unusual, and see no reason why it should persist.

I MISSED SEEING the original ice cores yesterday, but one industrial-scale freezer with backup generator is pretty much like another. Being summer, almost everyone is out at North GRIP, the new site whose cores might resolve the conflict between the two Greenland Summit cores about the cold-and-dry centuries during the last warm period 125,000 years ago.

This new site at 75°N 42°W is about 315 km north-northwest of the pair of summit sites, picked because radar soundings indicated that the Eemian layers ought to be well above the distortions caused by bedrock. They're using a big C-130 Hercules to fly in equipment and supplies. (They even hauled in a big road grader to groom the runway!)

I'd have enjoyed seeing the "cold hard data" but you can't actually see much in the ice cores without a lot of instrumentation. They measure the hydrogen and oxygen isotopes to infer air temperatures at the time the snow fell, and the dust particles give a nice indication of the dusty periods (much of the dust was kicked up far away, in the Gobi Desert, rather than from sources closer to Greenland). And they measure air bubbles trapped inside the ice, giving them a nice look at carbon dioxide, nitrous oxide, and methane concentrations back through prehistory, and how they co-vary with temperature.

There's a lot of useful information ("Data that makes a difference" is one definition of information) in one of those cores, whose cumulative length is about 3 km (Rich Alley refers to this as the "Two-Mile Time Machine"). Then comes the hard job of making organized bodies of knowledge out of mere information, creating a retrospectroscope. I spent a week listening to 80 paleoclimatoligists and climate modelers argue about the interpretation of the data from ice and sediment cores, how it eliminated some proposed explanations for what was driving the changes in temperature and rainfall, and how it suggested other possible explanations. From solid bodies of knowledge sometimes comes wisdom, so you realize what civilization should avoid and what we might do to help stabilize climate.

THE WHIPLASH CLIMATE CHANGES were easily one of the biggest scientific shocks of the last decade. You may not have heard of them, however, even though they were frequently mentioned in the news columns of the major scientific journals such as *Science* and *Nature,* with catchy titles such as "How ice age climate got the shakes." That's where the popular press usually picks up the important new science stories. What happened to delay the news getting out to general audiences about such an important story, and for more than ten years? The science reporters for the media were surely reading those *Science* and *Nature* summaries.

But the popular press has a pigeonhole labeled "Greenhouse," into which all climate change news is forced - and then either put down as "old news" or given the tired old "Has it *really* started yet?" treatment. Even worse is to assume the cooling will cancel out the warming, just as you turn up the cold water if the hot water supply suddenly improves in mid-shower. They have repeatedly missed the point that, rather than "averaging out," a gradual warming may trigger a dangerous cooling - a backlash. A climate full of whiplash coolings and warmings isn't a prediction but a newly-acquired historical perspective. What happened many times before is likely to happen again, unless we figure out a means of stabilizing the climate.

The data are very convincing. Deep in the ice sheets of Greenland are annual layers that record what the atmospheric gases and the air temperature were like over each of the last 250,000 years. That's the period of the last two major ice ages. A given year's snowfall is compacted into ice during the ensuing years, trapping air bubbles, and so paleoclimate researchers have been able to glimpse ancient climates in some detail. Water falling as snow on Greenland carries an isotopic "fingerprint" of what the temperature was like en route, thanks to oxygen-18 being a little heavier than the usual oxygen-16.

Counting those tree-ring-like layers in the ice cores shows that cooling came on abruptly. In the first few years the climate could cool as much as it did during the Little Ice Age, with tenfold greater changes over the next decade or two.

Though we tend to talk most about the temperature, the ice cores also reveal at lot about ancient dust, carbon dioxide, swamp-gas methane, and salt spray. In the layers from recent years, you find such anthropogenic gases as the chlorofluorocarbons used for aerosol spray cans, refrigerators, and foam containers. When winter storms have been violent, usually because of an increased contrast between the average temperature of land and ocean surfaces, you find a lot more dust and salt spray (indicates wind over oceans) in the cores.

They confirm the story told by the oxygen isotopes for temperature: things change very quickly. In just several years, every terrestrial mammal, including humans, was likely in big trouble from the droughts and the ensuing forest fires. Populations crash.

The onset of a cooling lasting centuries is as quick as a drought. Indeed it's the accompanying drought that we'd probably complain about first, along with unusually cold winters in Europe, were another climate flip to begin next year. The following summer, we'd complain about the smoke from the unrelenting forest fires, and about an even more severe crop failure. The sun would not so much set as disappear into a red haze above a murky horizon.

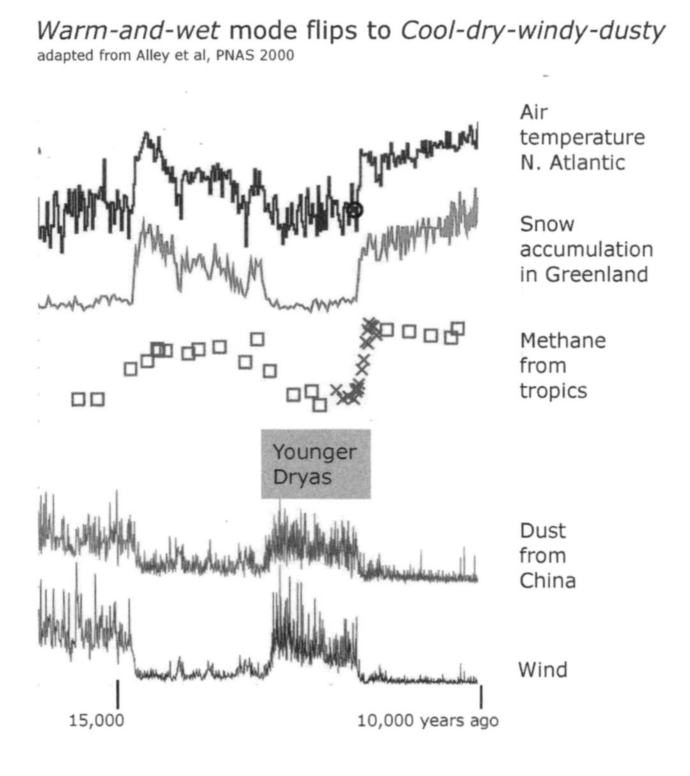

SO THAT'S THE "WHAT" of the ice ages, the major puzzles for which we need a "how" mechanism and a "why" long-term perspective that illuminates the predecessors which evolved into the present state of things. The how turns out to mostly be about salt, and how you get rid of buildups. The longest-term why concerns such things as the perpetual westerlies at the higher southern latitudes and continental drift moving Antarctica out of their way (the ring of westerlies probably started about 12 million years ago). The shorter-term why has to do with the Old Panama Canal getting dammed up. Maybe before the end of this nine-hour flight, I'll say something about "What next?" and the prospects for intervening to postpone the next abrupt climate change.

The flight from Copenhagen to Seattle is, in contrast, as timeless as heaven is reputed to be. We left about noon and we'll arrive about noon, too - having crossed nine time zones in nine hours of flying. (This plane's just the place to listen to a CD of Laurie Anderson singing about the little clock on your VCR, always blinking twelve noon because you never figured out how to get in there and change it.) A window seat in the stratosphere certainly provides a better place from which to contemplate the world than most philosophers ever had.

1985

NATURE VOL. 315 2 MAY 1985 REVIEW ARTICLE 21

Does the ocean–atmosphere system have more than one stable mode of operation?

Wallace S. Broecker*, Dorothy M. Peteet† & David Rind†

* Lamont-Doherty Geological Observatory of Columbia University, Palisades, New York 10964, USA
† NASA/Goddard Space Flight Center, Institute for Space Studies, 2880 Broadway, New York, New York 10025, USA

The climate record obtained from two long Greenland ice cores reveals several brief climate oscillations during glacial time. The most recent of these oscillations, also found in continental pollen records, has greatest impact in the area under the meteorological influence of the northern Atlantic, but none in the United States. This suggests that these oscillations are caused by fluctuations in the formation rate of deep water in the northern Atlantic. As the present production of deep water in this area is driven by an excess of evaporation over precipitation and continental runoff, atmospheric water transport may be an important element in climate change. Changes in the production rate of deep water in this sector of the ocean may push the climate system from one quasi-stable mode of operation to another.

To: Human Evolution E-Seminar
From: William H. Calvin
Location: 60.0°N 10.7°E 10,000m ASL
High above Oslo, Norway

Subject: **The ocean has a conveyor belt**

Glaciers descending from the north hereabouts lead most people to think that the center of the ice cap must have been the North Pole (that is, after all, the way it works at the South Pole). Wrong. It took a while after the discovery of the ice ages before anyone realized that glaciers don't form over open ocean: the pack ice at the North Pole is a few meters thick. Terra firma is four thousand meters farther down. The bottom of the Arctic Ocean is as deep as the Atlantic. It has features such as Nansen Basin, underwater ridges such as the Nansen Cordillera. They're both named for the Norwegian scientist Fridtjof Nansen, an early Arctic explorer (and, incidentally, one of the first neurobiologists).

As a naval officer sitting next to me on another airplane flight once remarked, "If anyone ever builds a house there, they'll get a surprise if they dig a basement!" (his submarine had punched through the ice, and he had gone walking on top of the world). But his trip was a few decades ago when the average sea ice depth was 3.1 meters; now it is down to 1.7 meters and models suggest that it will continue thinning and retreating with our global warming.

Although it may snow up on top, the floating ice seldom gets more than a few meters thick. To build up ice to the height of a mountain range, as happens during an ice age, requires a solid foundation such as Greenland. Down at the South Pole, there is a whole continent (9.3 percent of the Earth's land surface) to house tall piles of ice. When an ice age really gets going, then the northern hemisphere has a lot more land on which to house ice sheets than the southern. Typically, there have been big piles, housed on the continents either side of Greenland. Greenland itself still has a pile two miles high. The Canadian one created a 1000-meter-deep ice shelf pushing out into the Arctic on the continental shelf off northern Alaska.

The sea ice itself is of interest. It reflects 60 to 90 percent of the arriving sunlight back into space, thus keeping the earth somewhat cooler. At present, the warm Norwegian Current prevents a lot of sea ice. In the winter, one now sees sea ice along the eastern coast of Greenland above 70°N, but not along the Norwegian coast. Without that warm influence in the sub-Arctic, sea ice may come down past Norway to France - and that's a considerable percentage increase in whiteness, reflecting back summer sunlight that might help re-warm things. Like the self-perpetuating droughts (page 152), there are positive feedbacks for ice that make things even worse.

THERE ARE BIG SIGNS of civilization down below. Civilizations accumulate knowledge, so we now know a lot about what has been going on, what has made us what we are. We puzzle over oddities, such as the climate of Europe - which is far warmer than, by rights, it ought to be.

Oslo is anomalous, when you look around the globe for other major cities so far from the equator. In the southern hemisphere, latitude 60° is halfway between the tip of South America and Antarctica, down in the Drake Passage where strong westerly winds circle the world, stirring up prodigious wave heights. Though the long summer twilight of the higher latitudes may be nice, the flip side includes those dim winter days where the sun makes only a brief midday appearance low in the sky, barely edging the thermometer upward.

The populous parts of the United States and Canada are mostly between the latitudes of 30° and 45°, whereas the populous parts of Europe are ten to fifteen degrees farther north. "Southerly" Rome lies near the same latitude, 42°N, as "northerly" Chicago - and the most northerly major city in Asia is Beijing, near 40°N. London and Paris are close to the 49°N line that goes through Hudson Bay and, west of the Great Lakes, separates the United States from Canada. Berlin is up at 52°, Copenhagen and Moscow at about 56°. Oslo is nearly at 60°N, as are Stockholm, Helsinki, and St. Petersburg; continue due east and you'll encounter Anchorage.

Europe's climate, obviously, is not like that of North America or Asia at the same latitudes. For Europe to be as agriculturally productive as it is (it supports more than twice the population of the United States and Canada), all those cold, dry winds that blow eastward across the North Atlantic from Canada must somehow be warmed up. The job is done by warm water flowing north from the tropics, variously called the Gulf Stream and, when nearing Ireland, the North Atlantic Current. This warm water then flows up the Norwegian coast, with a westward branch warming Greenland's tip, at 60°N. It keeps northern Europe about 5-10°C warmer in the winter than comparable latitudes elsewhere - except, of course, when it fails.

THE WHIPLASH CLIMATE CHANGES are appallingly sudden and painful. And remember that they can happen even in the midst of warm climates like our present one. There was one cooling back during the previous warm period about 122,000 years ago that lasted for a dozen centuries before rewarming finally occurred. There was another abrupt cooling at 117,000 years, but with no recovery. That's how the last warm period ended - suddenly (ice per se returned more gradually). There have been dozens of whiplashes since then, but the warm flickers up out of ice age temperatures never lasted more than a few centuries, and were never as warm as now.

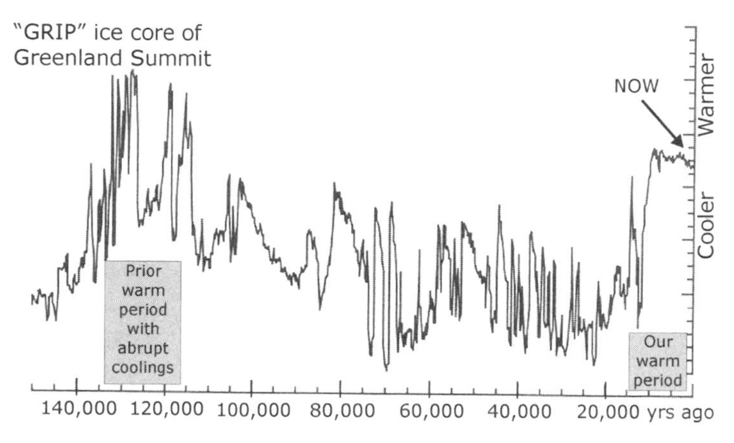

During the last big abrupt cooling, 12,900 years ago, Europe cooled down to Siberian temperatures within a decade (about ten-fold greater than in the Little Ice Age), the rainfall likely dropped by half, and fierce winter storms whipped a lot of dust into the atmosphere. Such conditions lasted for over 1,300 years, whereupon things warmed back up, even more suddenly. The dust settled and the warm rains returned, again within a decade.

Not only was Europe affected but also, to everyone's surprise, the rest of the habitable world appeared to be chilled about the same time. Tropical swamps decrease their production of methane at the same time that Europe cools, and the Gobi Desert whips much more dust into the air. When this happens something big, with worldwide connections, must be switching into a new mode of operation.

The North Atlantic Current is certainly something big, with the flow of about a hundred Amazon Rivers (an amount equal to all the rain falling on earth). And (to give you a little preview of the How, coming up) it sometimes changes its route dramatically, much as a bus route can be abbreviated into a shorter loop. Its effects are clearly global too, inasmuch as it is part of a long "salt conveyor" current that extends through the southern oceans into the Pacific.

Even a decade ago, we didn't know much about the climate flips; we simply thought that climate creep was starting to occur and that we needed to prevent greenhouse gases from slowly ramping up the heat. That too is still true, but we now know that the biggest threat from global warming is that it could trigger a far worse abrupt cooling, something akin to accidentally shifting into low gear when cruising at a high speed.

Declare the past, diagnose the present, foretell the future.
– HIPPOCRATES

I hope never to see a failure of the northernmost loop of the North Atlantic Current, because the result would be a population crash that would take much of civilization with it, all within a decade. Ways to postpone such a climatic shift are conceivable, however – old-fashioned dam-and-ditch construction in critical locations might even work. Although we can't do much about everyday weather, we nonetheless may be able to stabilize the climate enough to postpone an abrupt cooling. It all depends on developing some wisdom from all the new knowledge.

SUDDEN ONSET, SUDDEN RECOVERY – this is why I use the words "whiplash" and "flip-flop" to describe these climate changes. They are utterly unlike the changes that one would expect from accumulating CO_2 or the setting adrift of ice shelves from Antarctica. Change arising from some sources, such as volcanic eruptions or the West Antarctic Ice Sheet collapsing, may be abrupt – but they don't flip back just as quickly, centuries later.

Temperature records suggest that there is some grand mechanism underlying all of this, and that globally it has two major states, warm-and-wet and cool-and-dry-and-windy-and-dusty. In discussing the ice ages there is a tendency to think of warm as good – and therefore of warming as better. Alas, further warming might well kick us out of the warm-and-wet mode.

As for the How behind all the transitions, one naturally thinks of the sun first. The sun's variability, even though small enough so you'd think it insignificant, does correlate with some monsoon changes. But there is likely a chain of "causes," perhaps different for the abrupt warmings than for the abrupt coolings.

The likeliest reason for the abrupt coolings is an intermittent plumbing problem in the North Atlantic Ocean, one that seems to trigger a major rearrangement of the atmospheric circulation. North-south ocean currents help to redistribute equatorial heat into the temperate zones, supplementing the heat transfer by winds. When the warm currents penetrate farther than usual into the northern seas, they help to melt the sea ice that is reflecting a lot of sunlight back into space, and so the earth becomes warmer. Eventually that helps to melt ice sheets elsewhere. The major ice sheets take more than 10,000 years to disappear, a time scale a thousand times slower than some of the flip-flops in temperature.

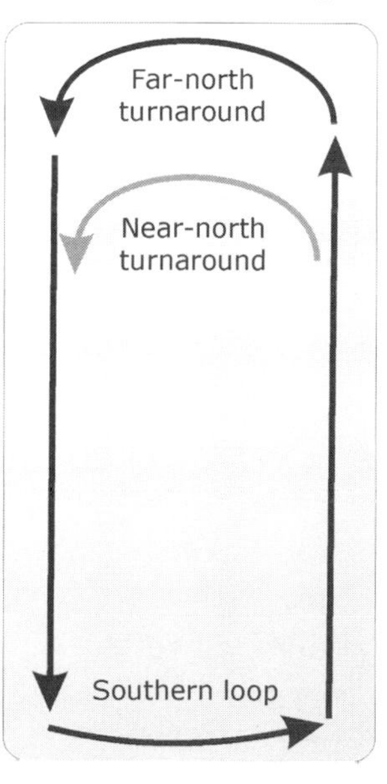

The warm-and-wet mode of global climate seems to involve ocean currents that deliver an extraordinary amount of heat to the vicinity of Iceland and Greenland. Like bus routes or conveyor belts, ocean currents must have a return loop. Unlike most ocean currents, the North Atlantic Current has a return loop that runs deep beneath the ocean surface. Huge amounts of seawater sink at known downwelling sites every winter, with the water heading south after it reaches the bottom. When that annual flushing fails for some years, the northern end of the usual conveyor belt stops moving and so heat stops flowing so far north – and apparently we're popped back into the cool-and-dry mode.

I have no trouble imagining this circuit because it is just like the Number 12 bus route in Seattle. A trolley bus heads north out of downtown and turns around near my home on northern Capitol Hill, returning to make a southerly loop through the downtown streets. But some Number 12 buses instead turn around at the bottom of Capitol Hill. Indeed, that near-north shortcut is where all of them turn around when ice closes the streets on Capitol Hill. Well, the warm-and-wet mode of climate corresponds to using the far-north turnaround and the cool-dry-dusty-windy mode to using only that near-north turnaround. (Imagine, if you like, the return path as taking the bus in a tunnel beneath the street.)

When ice puts a lid on the far-northern seas, the conveyor belt carrying heat north and salt south turns around well below Iceland, rather than in the far-north sinking sites near the northeast and southwest coasts of Greenland.

What if our climate jumped to something totally unexpected? What if you went to bed in slushy Chicago, but woke up with Atlanta's mild weather? Or worse, what if your weather jumped back and forth between that of Chicago and Atlanta: a few years cold, a few years hot? Such crazy climates would not doom humanity, but they could pose the most momentous physical challenge we have ever faced, with widespread crop failures and social disruption.

Large, rapid, and widespread climate changes were common on Earth for most of the time for which we have good records, but were absent during the few critical millennia when humans developed agriculture and industry. While our ancestors were spearing woolly mammoths and painting cave walls, the climate was wobbling wildly. A few centuries of warm, wet, calm climate alternated with a few centuries of cold, dry, windy weather. The climate jumped between cold and warm not over centuries, but in as little as a single year.

– RICHARD B. ALLEY, 2000

To:	Human Evolution E-Seminar
From:	William H. Calvin
Location:	63°N 6°E 10,000m ASL Out over the sinking Gulf Stream

Subject: **Dan's coffee cream trick**

We're leaving Europe's shoreline now, though it would have extended farther west back in the icy periods when the sea level was a lot lower because ice sheets locked up so much water. Except for the ice, you could have built a forty-story hotel at the seashore about 16,000 years ago; by 12,000 years ago, the waves would have been lapping at the windows halfway up. By 8,000 years ago, they'd have been over the top. A number of river valleys, carved in glacial times, were drowned when the sea level rose, such as the Thames below London, San Francisco Bay, and Chesapeake Bay.

The warm waters of the North Atlantic Current now flow up the west coast of Scotland and then continue all the way up the Norwegian coast; Oslo's harbor is ice-free all year. The warm current splits, some heading west toward Greenland, and the rest continuing north into the Greenland Sea, keeping a large area free of sea ice. I mentioned that sea ice can cap the ocean surface and prevent the winds from evaporating water and leaving salt behind. And that, along with rain and meltwaters, can interfere with sinking the surface waters. Let me warn you that the "how" of things is going to be pretty salty.

If you want to see cold sinking in action, pour some hot coffee into a tall transparent mug. Wait for a few minutes for the motion to settle. Then get some very cold cream (not half-and-half) and, using a spoon lowered to its rim in the coffee, gently pour the cream into the spoon, allowing it to overflow the rim and layer out onto the surface of the coffee. If you do this right, as my friend Dan Hartline showed me a decade ago, you will soon be rewarded by the sight of a column of cream plunging to the bottom. Though cream often floats because of its fat content, its density increases when cold, enough to sink through hot coffee. Indeed, the density buildup from salt excess and evaporative cooling is what causes the North Atlantic surface waters to sink so dramatically. (Unlike the thermohaline circulation, Dan's trick has a reversing feature: Once the cream warms down on the bottom, it will come geysering up to the top again in some very pretty turbulent plumes, finally ending up in a layer on top.)

Besides downwelling, you can create upwelling, by blowing gently from one side of the rim of the coffee cup. You'll push a wave of coffee across the surface, making room for deeper coffee to rise to the surface nearest your lips. Winds cause oceans to upwell in many places, such as off the north coast of South America.

SURFACE WATERS ARE FLUSHED regularly, even in lakes. Twice a year they sink, carrying their load of atmospheric gases downward. That's because water density changes with temperature.

Fresh water is densest at about 4°C (a typical refrigerator setting; anything that you take out of the refrigerator, whether you place it on the kitchen counter or move it to the freezer, is going to expand a little). A lake surface cooling down in the autumn will eventually sink into the less dense (because warmer) waters below, mixing things up. Seawater is more complicated, not expanding much below about 5°C but with the salt content becoming very important in determining whether water floats or sinks. Because surface water that evaporates leaves nearly all of its salt behind, the surface becomes saltier – and if it becomes more dense than the underlying water, it sinks, sometimes in great blobs that do not mix very well with underlying waters, just like Dan's cream.

The fact that excess salt is flushed from surface waters has global implications, some of them recognized two centuries ago. Salt circulates, because evaporation up north causes it to sink and be carried south by deep currents. That the cold waters of the ocean depths came from the Arctic was posited in 1797 by the Anglo-American physicist Sir Benjamin Thompson (the Count Rumford that I mentioned back in Germany), who also posited that, if merely to compensate, there would have to be a warmer northbound current as well. By 1961 the oceanographer Henry Stommel was beginning to worry that these warming currents might stop flowing if too much fresh water was added to the surface of the northern seas. By 1987 the geochemist Wallace Broecker was piecing together the paleoclimatic flip-flops with the salt-circulation story and warning that small nudges to our climate might produce "unpleasant surprises in the greenhouse."

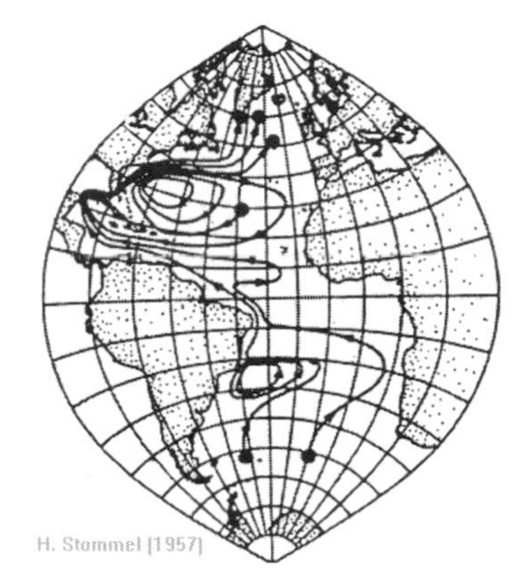

OCEANS ARE NOT WELL MIXED at any time. Like a half-beaten cake mix, with strands of egg still visible, the ocean has a lot of blobs and streams within it. When there has been a lot of evaporation, surface waters are saltier than usual. Sometimes they sink to considerable depths without much mixing, as happens in the eastern Mediterranean where the surface gets salty because evaporation exceeds the input from rivers. The salty bottom water flows west and out the bottom of the Strait of Gibraltar into the Atlantic Ocean. This water is about 10 percent saltier than the ocean's average, and so they sink into the depths of the Atlantic. A nice little Amazon-sized waterfall flows over the ridge that connects Spain with Morocco, 800 feet below the surface of the Strait.

Another underwater ridge line stretches from Greenland to Iceland and on to the Faeroe Islands and Scotland. It, too, has a salty waterfall, which pours the hypersaline bottom waters of the Greenland Sea and the Norwegian Sea south into the lower levels of the North Atlantic Ocean. This salty waterfall is more like thirty Amazon Rivers combined. Why has the eastern branch of it declined more than 20 percent in the last fifty years?

Indeed, why does it exist? The cold dry winds blowing eastward off Canada evaporate the surface waters of the North Atlantic Current, and leave behind all their salt. In late winter the heavy surface waters sink en masse. These blobs, pushed down by annual repetitions of these late-winter events, flow south, down near the bottom of the Atlantic. The same thing happens in the Labrador Sea between Canada and the southern tip of Greenland.

Salt sinking on such a grand scale in the Nordic Seas allows warm water to flow much farther north than it might otherwise do. It has been called the Nordic Seas heat pump. Nothing like this happens in the Pacific Ocean (which is, in consequence, about 5°C cooler), but the Pacific is nonetheless affected, because the sink in the Nordic Seas is part of a vast worldwide salt-conveyor belt. Such a conveyor is needed because the Atlantic is saltier than the Pacific (water which evaporates from the Atlantic is carried by the trade winds across Central America to fall as rain in the Pacific).

The Atlantic would be even saltier if it didn't mix with the Pacific, in long, loopy currents. These carry the North Atlantic's excess salt southward from the bottom of the Atlantic, down into the southern oceans, and some continues into the Pacific Ocean. A round trip on a grocery-checkout conveyor belt takes less than a minute. This conveyor takes more than a thousand years to make a complete loop.

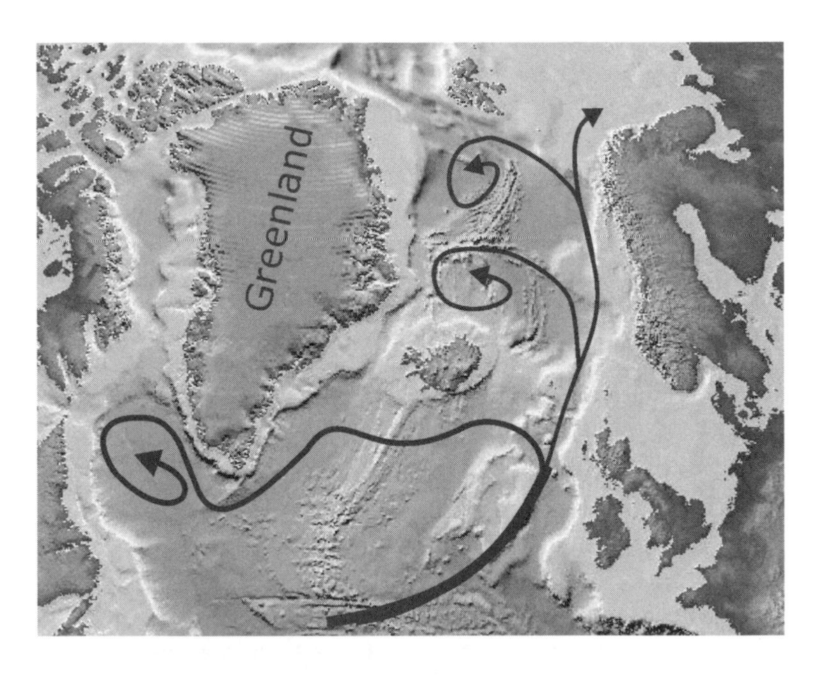

The floor of the North Atlantic Ocean showing the major far-north downwelling sites. The Gulf Stream can also sink at "near-north" sites near the bottom of the picture, especially when floating ice caps the far-northern sinks.

Everything else being equal, the Coriolis effect tends to make the major currents turn towards the right in the Northern Hemisphere. But shorelines and continental shelf walls may prevent right turns, as when the Norwegian Current continues north until attracted to the left by the sinking sites. Gyres are counterclockwise, however; as surface waters are attracted toward sinking sites, they will turn right.

Over the last million years, the pattern recorded in cores of Greenland ice has occurred over and over: a long stagger into an ice age, a faster stagger out of the ice age, a few millennia of stability, repeat. The current stable interval is among the longest in the record. Nature is thus likely to end our friendly climate, perhaps quite soon – the Little Ice Age may have been the first unsteady step down that path.

– RICHARD B. ALLEY, 2000

The climate record kept in ice and in sediment reveals that since the invention of agriculture some 8,000 years ago, climate has remained remarkably stable. By contrast, during the preceding 100,000 years, climate underwent frequent, very large, and often extremely abrupt shifts. Furthermore, these shifts occurred in lockstep across the globe. They seem to be telling us that Earth's climate system has several distinct and quite different modes of operation and that it can jump from one of these modes to another in a matter of a decade or two. So far, we know of only one element of the climate system which has multiple modes of operation: the oceans' thermohaline circulation. Numerous model simulations reveal that this ["conveyor"] circulation is quite sensitive to the freshwater budget in the high-latitude regions where deep waters form.

– WALLACE S. BROECKER, 1997

All climate models are predicting that [increasing greenhouse gases] will lead to a substantial temperature increase (around 2°C by the year 2100). The hydrological cycle of evaporation and precipitation is also expected to increase, as in a warmer world the atmosphere can hold more moisture.... A warmer climate will mean... more precipitation, possibly also extra freshwater from a melting of the Greenland ice sheet. The delicate balance in which the present conveyor operates may cease to exist. Model scenarios for the twenty-first century consistently predict a weakening of the conveyor by between 15 and 50 percent of its present strength.

– STEFAN RAHMSTORF, 2001

To:	Human Evolution E-Seminar
From:	William H. Calvin
Location:	71.0°N 8.9°W 10,000m ASL Jan Mayen Island

Subject: **Flushing the Gulf Stream**

Down below is Jan Mayen Island, featuring a volcano called Beeren Berg whose erect cone sweeps upward like the tip of Japan's Fujiyama. There's also a panhandle, a long ridge to the southwest of the volcanic cone. There's a Norwegian weather station down there, and not much else. But it would make a nice base for a dozen ships to monitor those downwellings year-round. The Greenland Gyre is just to the north, where the warm waters swing away from the ice-free Norwegian coast and head west, sinking here and there.

Besides the downwellings of the Greenland Sea, there are others in the central Labrador Sea between Canada and the southern tip of Greenland. Without such far-north downwellings, the world's climate is utterly different, far colder and drier in many important places around the world, though Europe takes the hardest hit.

Downwellings are hard to see, as only a few sink the surface waters via obvious whirlpools. River runners sometimes have to cope with downwellings in turbulent waters. If you see one side of your boat being sucked under, you shout "Highside!" and everyone throws their weight to the other side, trying to counterbalance. Waves that appear to flow beneath other waves are a tip-off that you're close to a downwelling. But those are little localized downwellings, and their cause is simply turbulence, usually from trying to force a square river into a round hole. The downwellings of the Greenland Sea and the Labrador Sea, usually in late winter, are likely much larger and even include giant 10-15 km wide whirlpools.

You know they've happened, however hard they are to see, because towing an instrument package on a long cable through the ocean depths will occasionally reveal a great blob of water with lots of dissolved oxygen and other atmospheric gases, quite unlike the surrounding waters in the depths. Oceanographers can follow a blob around for years. Remnants of the great 1988 downwelling in the western Labrador Sea were, eight years later, nearing Scotland at intermediate depths – it's a downwelling that wasn't salt-heavy enough to sink all of the way to the abyss.

"Almost all the way down" is what often happens to the northern end of the North Atlantic Current. The reason for the surface instability isn't just the cooling, though that helps. It's also because of all the evaporation promoted by the tropical heat when it reaches the high latitudes. The cold dry winds blowing eastward off Canada eagerly evaporate the warm surface water of the North Atlantic Current, and leave it heavy with excess salt. Eventually, late in the winter, the surface layer sinks in a big way. The heaviest blobs sink to the abyss and flow south, down near the bottom of the Atlantic.

What is so significant for our present climate is that such profound salt sinking makes room for warm water to flow much farther north than it might otherwise do. This "extension of the bus route" produces a 30 percent bonus of heat beyond that provided by direct sunlight to these seas, accounting for the mild winters downwind in Western Europe.

If it isn't obvious why salt sinking attracts warm water further north than otherwise, an analogy may help. A friend once discovered that the plumbers had inadvertently gotten their pipes crossed in a university building: the tank for a toilet was being refilled with hot water. This was not immediately apparent because the tank would ordinarily cool down to room temperature between flushes. But in periods of heavy use, it was noticed that the toilet tank was acting as a radiator, nicely warming up the room on cold winter days. We joked about inventing an automatic flushing device, triggered by the water cooling down to room temperature, so as to attract more hot water into the room.

That's an imperfect analogy for how the northern loop of the North Atlantic circulation is maintained and why Europe stays anomalously warm - but only so long as there is frequent flushing to replace the cooled Greenland Sea water with warm water attracted in from elsewhere.

Nothing like this happens in the Pacific Ocean. And sometimes failures of flushing occur in the North Atlantic.

1987

NATURE VOL. 328 9 JULY 1987 COMMENTARY 123

Unpleasant surprises in the greenhouse?

Wallace S. Broecker

There is now clear evidence that changes in the Earth's climate may be sudden rather than gradual. It is time to put research into the build-up of carbon dioxide in the atmosphere on a better footing.

THE inhabitants of planet Earth are quietly conducting a gigantic environmental experiment. So vast and so sweeping will be the consequences that, were it brought before any responsible council for approval, it would be firmly rejected. Yet it goes on with little interference from any jurisdiction or nation. The experiment in

hunches. They come from viewing the results of experiments nature has conducted on her own. The results of the most recent of them are well portrayed in polar ice, in ocean sediment and in bog mucks. What these records indicate is that Earth's climate does not respond to forcing in a smooth and gradual way. Rather, it

This record does not show the gradual change scientists had become accustomed to. Instead it shows an abrupt end to glacial time and, even more interesting, a brief period of intense cold interrupting the warm period that followed (Fig. 1). Although the two records shown in Fig. 1 are quite different, they are not incompat-

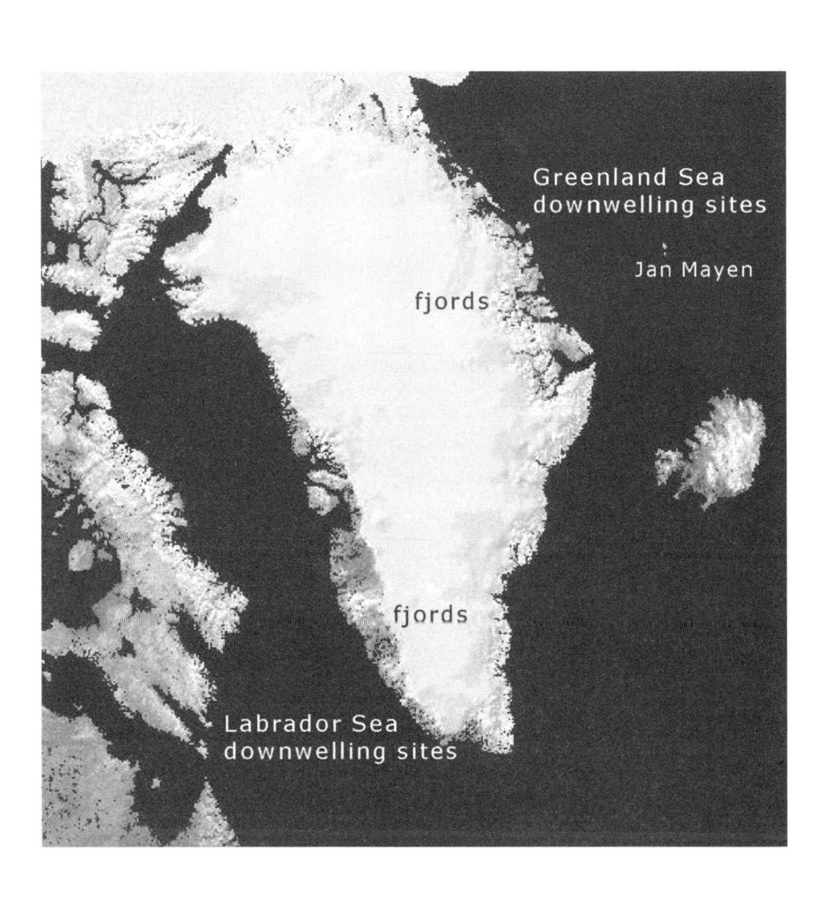
Greenland Sea
downwelling sites
Jan Mayen
fjords
fjords
Labrador Sea
downwelling sites

To: Human Evolution E-Seminar
From: William H. Calvin
Location: 72°N 12°W 10,000m ASL
The Greenland Sea

Subject: **Losing the first Panama Canal**

That's the what and where of our chattering climate - and a little of the short-term how. The sink in the Greenland Sea is simply part of a vast conveyor belt that carries the North Atlantic's excess salt to the less salty southern oceans and eventually to the Pacific Ocean. The most easily understood part of the return loop is where the conveyor rises again in the southern oceans. There is little land to get in the way of winds and wind-driven waves at the southern latitudes that correspond to Europe in the north; the tip of South America is about all there is, with a large gap until reaching the Antarctic peninsula. The winds reach in a complete circle, making for rough seas.

These westerly winds, thanks to the Coriolis effect, move the surface waters a bit to the left (the so-called Ekman current), and this allows waters from intermediate depths to rise to the surface. Models show that the southern westerlies alone are sufficient to start a conveyor running; what's added up in the far north (and in the narrow North Atlantic alone) is a highly efficient return loop of sinking water at the higher northern altitudes.

This up-elevator for deep water (the "Ekman suction") is the southern end of my analogy to Seattle's Number 12 bus route, where the bus comes out of the tunnel to venture north again on the surface streets. One of the far-north turnarounds is right beneath me here in the Greenland Sea. There's another major far-north turnaround in the Labrador Sea. The westbound branch from Ireland, that hooks around the southern end of Greenland, sinks off the southwest coast of Greenland. Sometimes most of the turnarounds occur east of Greenland but, a decade later, most of the action has shifted to the Labrador site.

All of the offshore Greenland sites qualify as far-north for purposes of my bus route analogy, but they fail when the winds are wrong or the floating ice caps them. The near-north turnarounds, corresponding to cool-and-dry climate, are in the oceans south of, say, 55°N where winter sea ice is less likely (see the picture on page 232 again).

The analogy to the Number 12 is incomplete because oceanographers imagine a third mode of global climate, seen only in the aftermath of Heinrich events, where the Hudson's Bay ice mountain collapses and iceberg armadas sail from Canada across to France, dropping rocks off their bottoms all the way (which is how they were detected, as layers in ocean-floor cores that get thicker and thicker as you get nearer Hudson Strait). As the icebergs melt along the way, they spread a lot of fresh water over the ocean surface. This may

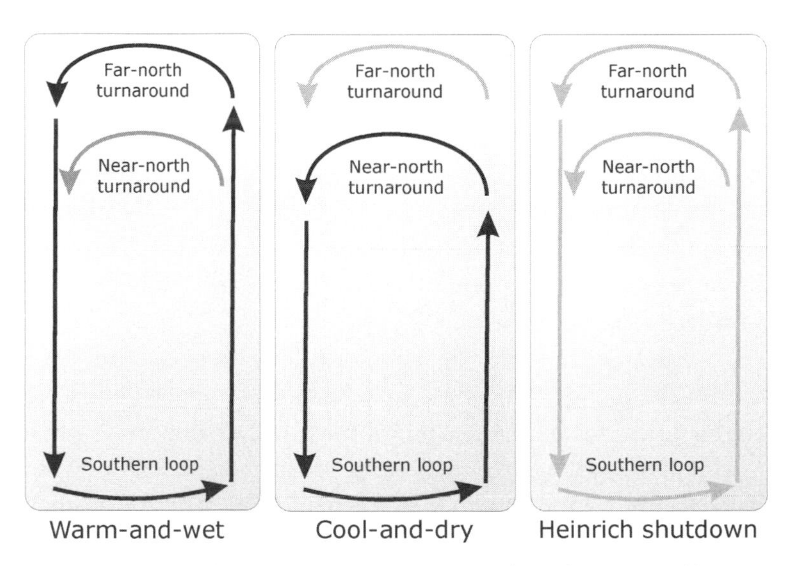

Warm-and-wet Cool-and-dry Heinrich shutdown

eliminate even the near-north turnaround, making it really

cold and dry up north. The far-southern oceans warmed somewhat during Heinrich events, what with the shutdown of the route for exporting heat.

Come to think of it, the Number 12 has a third mode too. The heavy snows from the 1996-1997 La Niña shut down both the far-north and near-north turnarounds of the Number 12, leaving the overheated buses accumulating downtown.

SO THE HEAT CONVEYOR is a process like that bus route, and not a particular place; it's a *how* that feels familiar to a physiologist like me. And the *why* of the conveyor belt is starting to fall into place, and it has to do with what the conveyor carries south: excess salt. The trade winds blow rain clouds across Panama, carrying Atlantic water to the Pacific without its salt. Without a means of exporting the Atlantic's excess salt to the southern oceans or the Pacific, the Atlantic's salinity would rise. While there is a diluted stream into the South Atlantic that comes around the tip of South America, the salinity problem is mostly handled by the conveyor, as it carries extra-salty waters south from the near-Greenland sinking sites. It's a heat-north/salt-south conveyor.

There used to be another major way of handling the salt buildup problem. If you've forgotten about that express route from Pacific to Atlantic located in the tropics, just think of it as the Panama Canal's predecessor. Continental drift connected North America to South America about three million years ago, damming up the easy route for disposing of excess salt and creating an instability via the salt buildup. The dam, known as the Isthmus of Panama, may have been what caused the ice ages to begin a short time later, simply because of the forced detour.

This major change in ocean circulation, along with a climate that had already been slowly cooling for millions of years, may be what led to ice accumulation most of the time – but also to climatic instability, with flips every few thousand years or so between warm-and-wet and cool-and-dry. The shallowing of the "Central American Seaway" probably started to change ocean circulation at least 4.6 million years ago, intensifying the Gulf Stream and bringing warm water to the high northern latitudes. The evaporative cooling of the Gulf Stream added a lot of moisture to the Northern Hemisphere. This intensification of the Gulf Stream may well have postponed the onset of the ice ages for awhile. When the whole earth gradually cooled a little more, this added moisture allowed ice mountains to build up around the northern Atlantic. While this Gulf Stream heat-and-moisture exchanger created a warm-and-wet mode, things can still flicker back into cool-and-dry in just a few years.

So a few million years ago was an instance of North Atlantic warming setting up a worldwide cooling. There is a more immediate prospect too, because there are well-known mechanisms whereby our current global warming could trigger a nasty flip to cool-and-dry.

FROM MY WINDOW ON THE WORLD, I keep expecting to see a fleet of oceanographic research ships down in the Greenland Sea, plumbing the depths with improved instruments, trying to figure out the *what-how-why* questions of this salty river. But oceanographers and paleoclimatologists do not have big budgets, and they have not been able to co-opt the naval ships of the NATO countries to tow additional instrument packages around and launch their telemetry buoys.

We are over the season's remnants of sea ice that extend down the northeast coast of Greenland for much of the year. They are being pushed out of the Arctic Ocean by a circulation pattern that starts over near Siberia; occasionally you'll see a Siberian tree washed ashore along this coast of Greenland. Hereabouts, the ice moves southward at about 15 km every day, carried by the far less salty Greenland Current.

I can now see land ahead, fingers of bare red-orange rock with waves crashing over them, the forerunners of Greenland. We have come to the far shore of the Greenland Sea. The fjords are ahead, and I need to look for bathtub rings with my binoculars.

Greenland fjord with floating sea ice

Greenland mountain peaks uncovered
by the thinning ice sheet

To: Human Evolution E-Seminar
From: William H. Calvin
Location: 74°N 19°W 10,000m ASL
Greenland fjords

Subject: **What stops the conveyor, flips climate**

Flying above the clouds often presents an interesting picture when there are mountains below. Out of the sea of undulating white clouds, mountain peaks stick up like islands. Greenland looks like that, even on a cloudless day - but the great white mass between the occasional punctuations is not a fluffy cloud layer but a massive ice sheet, miles deep. In places this frozen fresh water descends from the highlands in a wavy staircase, looking far more massive and magisterial than any alpine glacier.

Twenty thousand years ago a similar ice sheet lay atop the Baltic Sea and the land surrounding it. Another sat on Hudson Bay, and reached as far west as the foothills of the Rocky Mountains - where it pushed, head to head, against ice coming down from the Rockies. These northern ice sheets were almost as high as Greenland's mountains, obstacles sufficient to force even the jet stream to make a detour.

The ice ages provide a valuable perspective on how climates change, a view of the times in which our brief warm interlude is situated. Figuring out the mechanisms for ocean circulation modes depends on the evidence from paleoclimate, and on the stabilities seen via the great computer simulations. Most of the whiplash climate changes of the past were during icy periods that had ice sheets in Canada and Scandinavia.

Now only Greenland's ice remains, but the abrupt cooling in the last warm period shows that a flip can occur in situations much like the present one. Oceanographers are busy studying present-day failures of annual flushing, which give some perspective on the catastrophic failures of the past.

WHAT COULD POSSIBLY STOP the salty conveyor that brings tropical heat so much further north, and limits the formation of these ice sheets? It's not an idle question, and flushing failures are the obvious candidate. But, just as in medicine where we have to ask what the natural history of a disease is (natural ups and downs can fool you into thinking that a treatment is working when it's just happenstance), we need to know what the natural history of flushing failure is.

In the Labrador Sea, flushing failed during the 1970s, was strong again by 1990, and then declined. In the Greenland Sea over the 1980s salt sinking declined by 80 percent. Obviously, local failures can occur without catastrophe – it's a question of how often and how widespread the failures are, and whether they occur simultaneously in both the Greenland Sea and in the Labrador Sea. Large-scale flushing at both those sites is certainly a highly variable process, and perhaps a somewhat fragile one as well. And in the absence of a flushing mechanism to sink cooled surface waters and send them southward in the Atlantic, additional warm waters do not flow as far north to replenish the supply.

There are a few obvious precursors to flushing failure. One is diminished wind chill, when winds aren't as strong as usual, or as cold, or as dry – as is the case in the Labrador Sea during the negative phase of the North Atlantic Oscillation. This decade-scale shift in the atmospheric-circulation pattern over the North Atlantic, from the Azores to Greenland, strongly affects wintertime downwind. At the same time that the Labrador Sea gets a lessening of the strong winds that aid salt sinking, Europe gets particularly cold winters and diminished rainfall.

Another precursor is more floating ice than usual, which reduces the amount of ocean surface exposed to the winds, in turn reducing evaporation. The sea ice in the major flushing sites is only partly indigenous - much comes sailing down from farther north to replace what melts. Fresher water freezes more easily. Thus an ice lid is one possible candidate for what occasionally shuts down the highly efficient north-of-sixty-degrees portion of the Gulf Stream circulation. Arctic cold could thus cause a much more general worldwide cooling, abruptly.

Yet another precursor, as Henry Stommel suggested in 1961, would be the addition of fresh water to the ocean surface, diluting the salt-heavy surface waters before they became unstable enough to start sinking. Two mechanisms can be seen when piecing together the past: icebergs melting in the Heinrich events, and the freshwater floods seen just before the Younger Dryas and the 8,200 year event. But there is a third way to get enough fresh water layered on the ocean surface: more rain falling into the North Atlantic.

Yes, I know that it sounds like carrying coals to Newcastle. But, at least in certain regions such as the Greenland and Labrador Seas, rain falling into the ocean is a serious problem. Lots more rain falling is now what is predicted to happen from global warming. Even that which falls on land at higher latitudes tends to be carried toward the sea on north-flowing rivers, and so into the already somewhat less salty Arctic Ocean. It can stop salt flushing. Whereas the Norwegian coast stays ice-free now, an addition of fresh water would allow it to freeze more easily and so cap the oceans, preventing the wind-driven evaporation that makes the sea water dense enough to sink.

There is also a great deal of unsalted water in Greenland's glaciers, just uphill from the major salt sinks. The last time an abrupt cooling occurred was in the midst of global warming. Many ice sheets had already half melted, dumping a lot of fresh water into the ocean. The Greenland Sea got a lot of re-freshing from the east side of Greenland, making the out-of-the-Arctic Greenland Current even less salty. The Labrador Sea not only got the meltwater from Greenland's west coast but also from the Hudson's Strait flow that came off of the great mountain of ice which accumulated atop Hudson Bay and Labrador.

In the models of greenhouse warming, the gradual dilution of northern oceans by rain eventually slows the flushing until it is about 25 percent below normal. Further dilution and flushing fails. And that doesn't even consider the melting of the ice sheets (or additional perturbations like El Niño or the North Atlantic Oscillation). Meltwaters could steadily dilute the ocean surface, but years' worth of melting can also be dumped into the ocean in just one day.

There are already some indications that the Greenland Sea route for Gulf Stream warming is in trouble. A fifty-year trend of decline in the deep cold water in one branch of the conveyor's return loop has just been discovered, suggesting (if not compensated elsewhere) a decline in the warm water surface delivery of more than 20 percent.

The stories scientists tell are not simply bedtime tales. They place us in the world, and they can force us to alter the way we think and what we do.
– THOMAS LEVENSON, 1989

Persons living in this modern world who do not know the basic facts that determine their very existence, functioning, and surroundings are living in a dream world. Such persons are, in a very real sense, not sane. We [scientists] . . . should do what we can, or we shall be pushed out of the common culture. The lab remains our workplace, but it must not become our hiding place.
– GERALD HOLTON, 1996

To: Human Evolution E-Seminar
From: William H. Calvin
Location: 75°N 40°W 10,500m ASL
Atop Greenland

Subject: **Why melting can cause cooling**

Floods of fresh water might also prevent flushing, long before the average salinity was altered year-round. In the mid-1990s, we saw floods burst forth from the ice sheets of Iceland. The mid-Atlantic Ridge goes through Iceland, and there are a number of volcanos and hot springs. Some are under the ice sheets and, when they heat up, they melt ice near the bottom of the ice sheet. Eventually, when the pocket of meltwater becomes large enough, its roof may collapse and the steam plume alerts the local pilots. But even before then, the hot water may find its way downhill, melting the ice in its path, and finally burst forth as a sudden flood, sweeping down the valleys into the sea.

The fjords of Greenland, because they are occasionally dammed up, offer additional and even more dramatic examples of the possibilities for freshwater floods. Fjords are long, narrow canyons, little arms of the sea reaching many miles inland; they were scoured by great glaciers when the sea level was lower. Greenland's east coast has a profusion of fjords between 70°N and 76°N, including one that is the world's biggest. If blocked by ice dams, fjords make perfect reservoirs for meltwater.

Glaciers pushing out into the ocean usually break off in chunks. Whole sections of a glacier, lifted up by the tides, may snap off at the "hinge" and become icebergs. But sometimes a glacial surge will act like an avalanche that blocks a road, as happened when Alaska's Hubbard glacier surged into the base of the Y-shaped Russell fjord in May of 1986. Its snout ran into the opposite side, blocking the fjord with an ice dam. Any meltwater coming in behind the dam stayed there. At Russell Fjord, it was a serious matter for the seals remaining behind

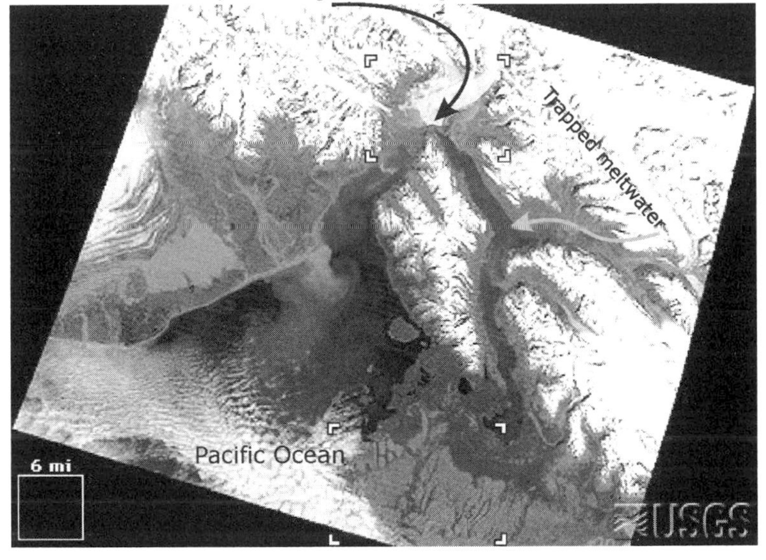

the dam, as they are used to salt water – and all the additional water coming in was unsalted, diluting the original sea water. A lake formed, rising higher and higher – up to the height of an eight-story building. That's why I look for bathtub rings and trimlines on the sides of Greenland fjords, just in case some floods of past centuries went unreported.

EVENTUALLY SUCH ICE DAMS BREAK, with spectacular results. Once the dam is breached, the rushing waters erode an ever wider and deeper path. Thus the entire lake can empty quickly. Five months after the ice dam at the Russell fjord formed, it started leaking about midnight and dumped a cubic mile of fresh water in the next twenty hours. Since the North Pacific Ocean lacks the downwelling so characteristic of the northernmost Atlantic, no great harm was done. Such an outburst flood from a fjord in Norway, Iceland, or Greenland might be a serious matter.

Worse, such a flood can break other ice dams in the vicinity. The Greenland fjords are long, with many side branches. As water rushes toward the sea, it also has a backwash up other branches, where it can weaken any ice dams there. So when warming conditions have produced glacial surges damming a number of nearby fjords, there is a domino effect: all of the meltwater reservoirs can be dumped within a few days. Were such a freshwater flood to occur from either the Greenland or Iceland coastlines, it might well prevent the flushing that makes room for more tropical water to flow northward.

Even ordinary flatland lakes sometimes find a new path to the sea, emptying over a century's time. The most recent abrupt cooling, a half-sized one about 8,200 years ago, appears to be due to a meltwater lake inland in Labrador about the size of present-day Lake Superior, which found a path into Hudson Bay, its waters thus coming into the Labrador Sea from Hudson Strait. This cooling was brief, with an exponential recovery over several centuries (unlike the usual sudden warming of the classic flip-flop events), and the flooding perhaps affected only half of the usual flushing sites (the Greenland Sea sites might have continued to flush the Gulf Stream). There was also a major outburst flood (from the huge meltwater Lake Agassiz) that came down the St. Lawrence, just before the Younger Dryas, 12,900 years ago.

We're nearing the northwest coast of Greenland. Somewhere south of here is that ex-lake in Labrador. But we're far north of it now; take a Copenhagen-Chicago flight if you want to see the scene of the most recent major climate crasher. You wouldn't think that rain falling in the ocean – nice, clean drinkable water – could cause so much trouble. Of course, no one thought that water would turn out to be comprised of two gases, either.

Wallace Broecker is worried about the world's health. Not so much about the fever of global warming but about a sudden chill. For more than a decade, the marine geochemist has been fretting over the possibility that a world warming in a strengthening greenhouse might suffer a heart attack, of sorts: a sudden failure to pump vital heat-carrying fluids to remote corners of Earth. If greenhouse warming shut down the globe-girdling current that sweeps heat into the northern North Atlantic Ocean, he fears, much of Eurasia could within years be plunged into a deep chill.

– RICHARD A. KERR, 1998

To:	Human Evolution E-Seminar
From:	William H. Calvin
Location:	78°N 69°W 10,400m ASL Thule, Greenland
Subject:	**Rube Goldberg chains of cause-and-effect**

Speaking of expensive undertakings of the past, I see one far below, though it hardly qualifies as a pyramid or a cathedral. It's the first sign of civilization since that lonely weather station on Jan Mayen Island. We are now over the fjord system of northwest Greenland, just 950 miles south of the North Pole, and there's a big airport below, considerably larger than anything that the Eskimo lifestyle requires.

It is surely Thule, that far-north U.S. air force base built in the summer of 1951 to refuel long-range bombers in the scary days of Joseph Stalin. American air force personnel out of favor with their superiors worried about being assigned to Thule, just as their Cold War counterparts worried about being sent to Siberia. Thule is well above the Arctic Circle and the sun simply doesn't rise for long months.

The most detailed evidence for climate change once came from Camp Century, a research camp atop the ice cap near Thule. Ice cores from there in 1966, and others from the Dye 3 site in southern Greenland in 1981, suggested some abrupt changes but various objections were raised to parts of the data. So two independent deep cores were drilled between 1989 and 1993 starting at the highest part of Greenland's ice cap down south, where the sideways movement of ice was minimal and unlikely to contaminate the chronology (the "GISP2" American team led by Pieter Grootes drilled 30 km from the "GRIP" European team led by Willi Dansgaard). And this way, they also got the longest possible record, striking bedrock after seeing ice that was 250,000 years old, the date of the warm period before last, the one that melted much of Greenland's ice sheet.

Having two nearby cores has allowed confidence in the interpretation of the last 115,000 years, though the lack of agreement between them during the last warm period 130,000 to 117,000 years ago has raised some questions about how much fluctuation there was in that warm period. There is little doubt that there was at least one "mid-Eemian cooling event" about 122,000 years ago, because it is seen in ocean-floor cores, coral reefs, and the pollen cores of European lakes. It lasted perhaps fifteen centuries before recovering to temperatures even warmer than today's.

But only the ice core methods have the resolution to say if there were other still-briefer events – and so the paleoclimate researchers are now looking at sites in North Greenland and Antarctica that might be capable of yielding undisturbed cores. As I mentioned back in Copenhagen, they're now drilling North GRIP. It's particularly important data to get, as you want to know how the whiplashes work in a warm period like today's, when Greenland and Iceland are the main source of North Atlantic meltwater and the sea level is at a modern height.

Speaking of meltwater, the results of floods may linger for decades, simply because the ocean waters mix so slowly. Nothing on this multidecade scale happens in the atmospheric circulation. But the oceans are sluggish enough so that we might say that they have a "memory."

THE GREAT SALINITY ANOMALY, a pool of water less salty than normal, was tracked moving around the North Atlantic between 1968 and 1982. It probably started in the Arctic Ocean, but it nicely illustrates what would happen if a fjord dam collapsed as it is derived from about 500 times as much unsalted water as that released by the four-month accumulation at Russell Fjord.

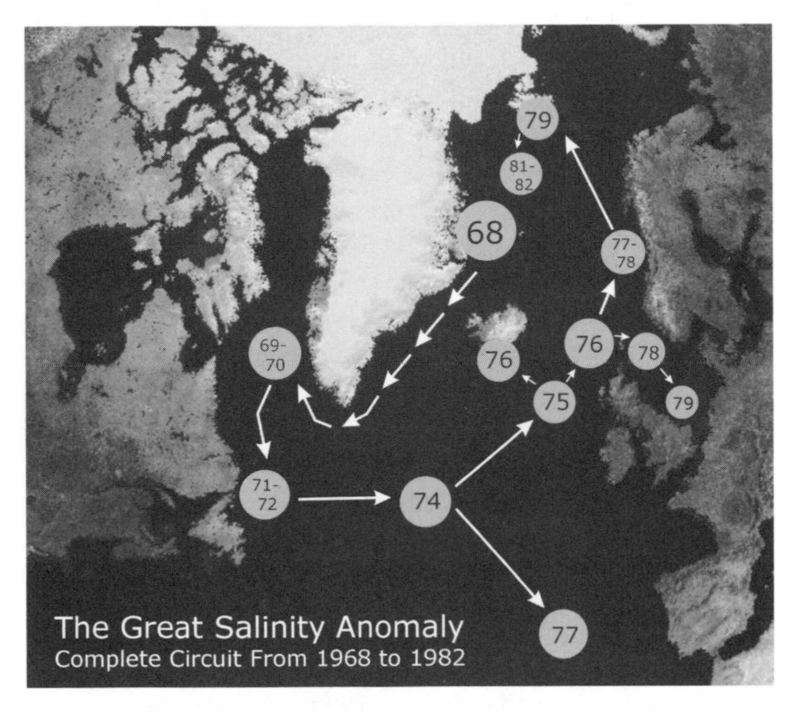

Two years after it was detected off Greenland's east coast, it arrived in the Labrador Sea, where it prevented the usual salt sinking. By 1971-1972 the semi-salty blob was off Newfoundland. It then crossed the Atlantic and passed near the Shetland Islands around 1976. From there it was carried northward by the warm Norwegian Current, whereupon some of it swung west again to arrive off Greenland's east coast – where it had started its inch-per-second journey. This counter-clockwise "sub-polar gyre" thus has a loop time of more than a dozen years.

So freshwater blobs drift, sometimes causing major trouble, and a Greenland flood thus has the potential for some years afterward to stop the enormous heat transfer that keeps the North Atlantic Current going strong. It shows us the lingering effects of a freshwater excess from any source. They circulate, causing trouble in a way that any engine mechanic can instantly appreciate, from water blobs in the fuel line.

There is potentially a longer version of this as well, since the conveyor belt takes about a thousand years to cycle.

Remember that fresh water freezes more easily than the ocean's usual salt water, so if downwelling fails locally, a puddle of fresher water may form from the rains or floods – and it will freeze more easily, preventing the winds from doing their evaporation job that might restart the downwelling. Once downwelling stops over a large area, it is hard to restart. Sea ice may form, stopping the winds from stirring the surface, stopping evaporation, leaving the passing winds as cool and dry as when they left Canada.

THE ABRUPT COOLINGS OF THE PAST may have been triggered by various causes at various times, and I will devote the next several pages to sketching them out for the fans of causation. Causes in the real world tend to come in layers, and there is an underlying 1,500-year cycle in North Atlantic ice rafting that goes back 100,000 years, in both our warm period and the antecedent ice age. The isotope record of the last 12,000 years of solar activity parallels it, suggesting a solar pacemaker. Stephen Porter sees the 1,500-year cycle in the layers of wind-blown silts in China called loess, even in warm periods like today. The latest episode in the 1,500-year cycle was probably the Medieval Warm Period that started about 1,200 years ago, dipping down into the Little Ice Age, and now warming again – though augmented by the extraordinary twentieth-century carbon dioxide rise, now well outside the bounds of carbon dioxide variation during the prior three ice ages.

This minor solar cycle may run all the time. But its effects may be amplified by sea salt. Waves bury air and when bubbles rise to the surface, they pop, spraying salt into the atmosphere. The salt ions provide nuclei for condensation, augmenting the number of water droplets that form in the air from the water vapor. This changes the brightness of the atmosphere (many small droplets reflect more than a few larger droplets made from the same amount of water vapor).

From this, you get more clouds. Sunlight returned to space doesn't heat the earth.

The cool-and-dry periods have a lot more of these airborne sodium and chloride ions. High winds not only produce more salt spray, but much of the dust falling on Greenland came from near large salt flats in China. So there is no lack of windy sources for elevating salt.

Salt is involved in another way as well: there ought to be a slow buildup of salinity in the Atlantic Ocean (the Atlantic exports more fresh water as rain than it gets back from rivers returning to the Atlantic) - unless, of course, long loopy currents between Atlantic and Pacific Oceans serve to even things out.

The geochemist Wallace Broecker, to whom we owe a number of the important ideas about abrupt climate change, speculates that there is a chain of causation starting with more far-northern winter sea ice and (because of the ice preventing the winds from stirring up waves and evaporation and salt excess) thereby fewer sinks for the Gulf Stream, which in turn diminishes the big conveyor loop of currents linking the North Atlantic to the Pacific. This decreases the export of excess salt to the Pacific; it also makes the high latitudes colder and the tropics warmer. This temperature contrast makes for more high intensity storms, which leads to more atmospheric dust and sea salts, which reflects more sunlight back into space and cools the earth further. As salt builds up in the Atlantic, it makes it easier to restart the far-northern downwellings.

So this system could oscillate, producing a slow cycle in Atlantic salinity and far-northern ice, even in warm-and-wet times like today. But much stronger albedo effects (a measure of how much sunlight is simply reflected back out into space) might be generated by the high winds of the glacial era, giving $10^{\circ C}$ temperature changes rather than the $1^{\circ C}$ excursion of the Little Ice Age. Since the 1,500-year period remains relatively unchanged, whether in an ice age or warm period, the sun's variability - though small - is an attractive explanation. Many people are looking for what amplifies the sun's effects on earth climate.

Something like albedo might explain the 1,500-year cycle without a two-state mechanism; the D-O flips might arise from an abrupt atmospheric reorganization triggered by accumulating regional differences in sea surface temperatures. (Anyone who works on nerve and muscle excitability will immediately recognize an analogy: we often see a modulation of the amplitude of analog processes so that they occasionally cross a threshold to trigger a flip.) I like it, and it really doesn't conflict with Broecker's earlier idea that atmospheric water vapor levels changed (presumably secondary to an atmospheric cellular circulation change). A one-two punch of albedo enhancement followed by less tropical evaporation might carry things from a warm-and-wet regime to a cool-and-dry mode.

What, someone asked by e-mail, causes carbon dioxide to decrease during the cold-and-dry periods? Dust storms. While weathering of rock is a long-term answer, it's probably not what causes the variations within an ice age. But soil, liberated by vegetation failures in the cool-and-dry mode's droughts, is carried by the winds into the ocean, fertilizing the plankton and so a whole food chain. A lot of carbon falls to the ocean floor in the form of fecal pellets, and so much carbon dioxide is taken out of circulation for a while.

IF THIS SEEMS LIKE A LOT OF "CAUSES," it is. In biology, we have even longer Rube Goldberg chains of cause-and-effect. In medicine, this gives us multiple opportunities to interrupt a chain of disease causation.

The staged "causes" of an abrupt cooling are going to include proximate causes - the *coup de grace* to the old climate might come from too much rain in the wrong place from a particularly severe El Niño finally tripping an atmospheric cell reorganization. But some near-proximate cause might set the stage, say the loss of crucial whirlpools that sink surface waters with high efficiency. That in turn might be set up by the albedo or whatever else amplifies the 1,500-year cycle, thus poising the climate near the tipping point, setting things up for a quick slide. Underlying that would be whatever creates the 1,500-year cycles (maybe Broecker's salt oscillator, maybe something else). And you'd expect even more ultimate "causes" such as continental drift making the North Atlantic narrow enough that the Norwegian Current is "boxed in," unable to turn right from the Coriolis effect, so it goes even farther north, hugging Europe's continental shelf - or uplifting Panama 3 million years ago to close off the old tropical route for equalizing salinity.

A greenhouse warming may reduce the 1,500-year cycle aspects, but this provides us with no comfort regarding future prospects since a warming can shortcut the usual circuit, bypassing the usual stage-setting by amplification of the 1,500-year cycle. No one is saying that greenhouse warming was the typical cause of climate flips - but clearly Greenland still has the fresh water supply that is capable of suppressing flushing. Clearly some of Greenland's fresh water will be released by any further warming of our climate, adding to the sea surface dilution produced by global warming's increased rainfall.

This assessment hinges on several things. First, that the oceanographers are correct in their present hypothesis, that a persistent failure of late winter sinking of the ocean surface near Greenland and Iceland is a likely cause of most of the abrupt cooling episodes. And second, that massive continental ice sheets (as existed over Canada and Scandinavia, when most of the abrupt coolings occurred) aren't needed to provide the necessary dilution of the salty surface - that rain or Greenland's meltwater will do, as in that abrupt cooling of the last warm period which occurred when the two continental ice sheets were long gone.

Let me try to summarize the state of affairs before I speculate a little, based on what I know from somewhat analogous nonlinear systems seen in biophysics and neurobiology that also flip.

WE HAVE COME A LONG WAY from the chimplike ancestor 5-6 million years ago. And we've come a long way in this e-seminar, up through our ancestral climates and changing ways of making a living. We've seen both the gradual and the abrupt. Where possible, I've tried to sneak in some of the mechanistic aspects, whether vertical curtains of air rising into the heavens or salt-heavy cold water sinking to the abyss.

If one is tugging on the dragon's tail with little notion of how much agitation is required to wake him, one must be prepared for the unexpected.
– RAY T. PIERREHUMBERT, 2000

But we're now up to the present, and wondering what the future will bring. I'll try to break the news gently, but what it boils down to is that the future is not an exact science. Neither is medicine. You're always having to act on incomplete data, because the failure to act can prove fatal.

Of this much we're sure: global climate flip-flops have frequently happened in the past, and they're likely to happen again. It's also clear that sufficient global warming could trigger an abrupt cooling in at least two ways - by increasing high-latitude rainfall or by melting Greenland's ice, both of which could put enough fresh water onto the ocean surface to suppress flushing. To that you can probably add a third, capping evaporation with sea ice.

Further investigation might lead to revisions in such mechanistic explanations, but the result of adding fresh water to the ocean surface is pretty standard physics. In the four decades of subsequent research, Henry Stommel's theory has only been enhanced, not seriously challenged.

Up to this point in the abrupt climate change story, none of the broad conclusions is particularly speculative. But to address how all these nonlinear mechanisms fit together - and what we might do to stabilize the climate - will require some speculation.

To:	Human Evolution E-Seminar
From:	William H. Calvin
Location:	73°N 95°W 10,400m ASL Somerset Island, NW Territories

Subject: **North Poles aren't what they used to be**

If this were 1831, the year that the North Magnetic Pole was first located on the Boothia Peninsula, it would be off to the left of the plane. But it has moved since then, and is now off to the right, about a thousand kilometers to the northwest of where it was in 1831.

The North Magnetic Pole is the point toward which all those decelerating charged particles from the solar wind converge to cause the aurora borealis. From space, the aurora looks like a fountain, spewing light - making the Magnetic Pole look considerably more exciting than the Geographical North Pole.

No northern lights for us today. They're there 24 hours a day, but it's still noon and we can't see them for all the summer sunlight that selectively scatters off the thin air to produce a blue sky. Now, in the winter, when it's most dark up here, you stand a better chance of seeing the northern lights converging on the magnetic pole.

And the Geographic North Pole isn't what it used to be, either.

BACK ABOUT FIFTY MILLION YEARS AGO, there was no sea ice at the North Pole, just open water like all the rest of the deep oceans. But more recently, Arctic explorers have always been able to plant a flagpole at the North Pole.

In the first summer of the new millennium, however, there was no place to stand at the North Pole - unless you stood up in a boat at 90°N. (These days, one holds aloft for the camera that flag of high-tech – a little yellow GPS unit reading a latitude of 89.99999°N and with the longitude wildly varying, each time the boat rocks.)

Radar ice-thickness estimates of the Arctic Sea ice showed that it had been thinning for years, just as they had also shown that the northern coastal glaciers of Greenland were thinning. But few guessed that you'd be able to take a Russian icebreaker all the way to 90°N and find open water in the summer, with no ice to stand on.

That it can happen at all is, of course, thanks to the Norwegian Current carrying water from the North Atlantic Current up to the major downwelling whirlpools at 76°N. But that it is so much warmer at 90°N in recent decades is likely due to things warming up more generally, bringing greenhouse predictions uncomfortably to mind.

Some scientists argue that the instrumental record (weather balloons and such) is equivocal on warming in the twentieth century. Given all of the ice reduction data from recent decades, this just tells you that we've been measuring temperature in the wrong places, or weighting the data wrong. Nature, obviously, is sensitive to some other set of temperatures than the ones we've been measuring - and Nature is, generally, what counts. To cite the twentieth-century instrumental record, without mentioning all the melting - as even the occasional scientist may do before general audiences - is often suspected of being special pleading, trying to confuse nonscientists with carefully selected facts, what lawyers with a losing case sometimes attempt to do with juries. The radar measurements in the Arctic tell the warming story in a way that will be hard for the global-warming skeptics to refute.

There's no place left to stand at the North Pole. Of course, the open-ocean gap, a fissure in the sea ice called a polynya, will move around, temporarily restoring some footing at exactly 90°N, but the point remains: Beware of thin ice, and its implications for the world to the south of 90°N.

YOU CAN SEE A MODE OF OPERATION in the Sahara as the earth's axis gradually tilts to change the penetration of the monsoons into arid areas. Just within decades, less dust blows offshore because the Sahara got major amounts of grass: small additional changes create major consequences. This started about 14,000 years ago. The tilt peaked about 9,500 years ago when there were lakes all around the Sahara, and then the arid conditions abruptly reappeared about 5,000 years ago when the declining tilt again passed through about the same value.

As with the self-perpetuating drought cycle (page 152), establishing (and disestablishing) vegetation is thought to have a lot to do with creating *regional* modes of climate. The big problem is in figuring out what can cause *worldwide* modes. There are some ideas, such as Gulf Stream failure, for how we might flip suddenly from warm-and-wet into cool-and-dry in a matter of decades. But sudden warming is the real puzzle; no one yet has a good idea for how things can flip back, and even more quickly.

> **Climate in the past has been wildly variable, with larger, faster changes than anything industrial or agricultural humans have ever faced . . . . Climate can be rather stable if nothing is causing it to change, but when the climate is "pushed" or forced to change, it often jumps suddenly to very different conditions, rather than changing gradually.**
> — RICHARD B. ALLEY, 2000

Even the tropics cool down by about 5°C during an abrupt cooling, and it is hard to imagine what in the past could have disturbed the whole earth's climate on this scale. We must look at arriving sunlight and departing light and heat, not merely regional shifts on earth, to account for changes in the temperature balance. Increasing amounts of sea ice and clouds could reflect more sunlight back into space, but Wally Broecker suggests that a major greenhouse gas is disturbed by the far-north failure of the salt conveyor, and that this affects the amount of heat retained.

In Broecker's view, failures of salt flushing cause a worldwide rearrangement of ocean currents, resulting in – and this is the speculative part – less evaporation from the tropics. That, in turn, makes the air drier. Because water vapor is the most plentiful greenhouse gas, this decrease in average humidity would cool things globally. As Broecker has said, "If you wanted to cool the planet by 5°C and could magically alter the water vapor content of the atmosphere, a 30 percent decrease would do the job."

Just as an El Niño produces a hotter Equator in the Pacific Ocean and generates more atmospheric convection, so there might be a subnormal mode that decreases heat, convection, and evaporation. In reconfiguring three cells per hemisphere into some other mode of general circulation, it might incidentally reduce the amount of tropical evaporation and thus shift us into a subnormal amount of greenhouse warming. (Be careful what you wish for!)

TO SEE HOW OCEAN CIRCULATION might affect greenhouse gases we must try to account quantitatively for important nonlinearities, ones that allow little nudges to provoke great responses – like the typical on-off light switch. Our usual gradualist extrapolations of the present state of affairs are more like dimmer switches. All metaphors break down somewhere, and the gradualist scenarios seem particularly likely to fail. Let me try some nonlinear metaphors to better approximate the climate mechanisms.

The modern world is full of objects and systems that exhibit bistable modes, with thresholds for flipping. Door latches suddenly give way. A gentle pull on a trigger may be ineffective, but there comes a pressure that will suddenly fire the gun. Household thermostats tend to activate heating or cooling mechanisms abruptly – also an example of a system that pushes back.

We must be careful not to think of an abrupt cooling in response to global warming as just another self-regulatory device, a control system for cooling things down when it gets too hot. The scale of the response will be far beyond the bounds of regulation - more like when excess warming triggers fire extinguishers in the ceiling, ruining the contents of the room while cooling them down. Though combating global warming is obviously on the agenda for preventing a cold flip, we could easily be blindsided by stability problems if we allow global warming per se to remain the main focus of our climate-change efforts.

Multiple consequences of a single cause are something we can think about, if reminded. ("You can't do just *one* thing.") What's far harder for us is to think about multiple causes at the same time. We can think about one cause and its most obvious consequences, but factoring in a few more simultaneously-acting causes usually requires much effort. Even experienced scientists and historians find it challenging. The politicians and press who read only "executive summaries" of climate change reports routinely oversimplify what they read, and the public gets even less of whatever wisdom was originally there. And since nothing expensive gets done unless the politicians feel they have the people behind them, the world's largest democracies may fail to act in time.

We are in a raft, gliding down a river, toward a waterfall. We have a map but are uncertain of our location and hence are unsure of the distance to the waterfall. Some of us are getting nervous and wish to land immediately; others insist that we can continue safely for several more hours. A few are enjoying the ride so much that they deny that there is any imminent danger although the map clearly shows a waterfall. A debate ensues but even though the accelerating currents make it increasingly difficult to land safely, we fail to agree on an appropriate time to leave the river. How do we avoid a disaster?

To decide on appropriate action we have to address two questions: How far is the waterfall, and when should we get out of the water? The first is a scientific question; the second is not. The first question, in principle, has a definite, unambiguous answer. The second, which in effect is a political question, requires compromises.

– GEORGE PHILANDER, *Is the Temperature Rising?*, 1998

To: Human Evolution E-Seminar
From: William H. Calvin
Location: 68°N 105°W 10,400m ASL
Crossing the coastline of North America

Subject: **How we might stabilize climate**

Predicting the future is sometimes possible on the short time scales of weather forecasting, but sensitive dependence on initial conditions makes long-term predictions iffy. Spotting trends, and acting to head off disasters, is one of the ways in which human intelligence improves upon mere physics. So our stories about possible futures are not predictions so much as scenarios.

Our attitude toward climate change has long been like that of the driver who cannot stop when his headlights illuminate an obstacle, because of going too fast. Science is providing us with somewhat better headlights. Indeed, it has just told us that global warming has an evil twin, that there are going to be some missing bridges on that dark road ahead, not just some bumps but some voids where the comfortable road drops out from under us.

And it isn't just a matter of slowing down (though that would help). We are near the end of a warm period in any event; ice ages return even without human influences on climate. The last warm period abruptly terminated 13,000 years after the abrupt warming that initiated it 130,000 years ago, and we've already gone 15,000 years from a similar warm-up starting point. But we may be able to do something to delay an abrupt cooling.

DO SOMETHING? This tends to stagger the imagination, immediately conjuring up visions of terraforming on a science-fiction scale – and so some shake their heads and say, "Better to fight global warming by consuming less," and so forth.

Surprisingly, it may prove possible to prevent flip-flops in the climate – even by means of low-tech schemes. Keeping the present climate from falling back into the cool-and-dry mode will in any case be a lot easier than trying to reverse such a change after it has occurred.

Were fjord floods causing flushing to fail, because the downwelling sites were fairly close to the fjords, it is obvious that we could solve the problem. All we would need to do is, using highway-construction amounts of explosives, open a channel through the dam – and do it before dangerous levels of fresh water build up.

This works only if *floods* of fresh water prove to be the problem, because the downwelling sites turn out to be close enough to the fjords. The jury's still out on that one. Timing could be everything, given the delayed effects from inch-per-second circulation patterns, but that, too, potentially has a low-tech solution: build dams across the major fjord systems and hold back the meltwater at critical times. Or divert eastern-Greenland meltwater to the less sensitive north and west coasts.

But relying on such simple fixes presumes that you know what you're doing. You get to be an expert in this field only by knowing the data and knowing the processes, backward and forward. And I mean that literally: you have to have computer models that successfully predict the past before you'll even think of trusting them to predict the next fifty years. Before we become as busy as beavers, we'll want to try out the dam schemes on the computer models. That's the way you find out the common mistakes, not by experimenting directly on our one and only global climate.

Fortunately, big parallel computers have proved useful for both global climate modeling and detailed modeling of ocean circulation. They even show the flips. Computer models might not yet be able to predict what will happen if we tamper with downwelling sites, but this problem doesn't seem insoluble. We need more well-trained people, bigger computers, more coring of the ocean floor and silted-up lakes, more ships to drag instrument packages through the depths, more instrumented buoys to study critical sites in detail, more satellites measuring regional variations in the sea surface, and similarly for studying the atmosphere. Eventually you'd progress to some small-scale trial runs of interventions.

It would be especially nice to see another dozen major groups of scientists doing climate simulations, discovering the intervention mistakes as quickly as possible and learning from them. Medieval cathedral builders learned from their design mistakes over the centuries, and their undertakings were a far larger drain on the economic resources and people power of their day than anything yet discussed for stabilizing the climate in the twenty-first century. We may not have centuries to spare, but any economy in which two percent of the population produces all the food, as is the case in the United States today, has lots of resources and many options for reordering priorities.

An expert is someone who knows some of the worst mistakes that can be made in his subject, and how to avoid them.
– WERNER HEISENBERG, 1971

FUTURISTS HAVE LEARNED to bracket the future with alternative scenarios, each of which captures important features that cluster together, each of which is compact enough to be seen as a narrative on a human scale. Three scenarios for the next climatic phase might be called population crash, cheap fix, and muddling through.

The population-crash scenario is surely the most appalling. Plummeting crop yields will cause some powerful countries to try to take over their neighbors or distant lands – if only because their armies, unpaid and lacking food, will go marauding, both at home and across the borders. The better-organized countries will attempt to use their armies, before they fall apart entirely, to take over countries with significant remaining resources, driving out or starving their inhabitants if not using modern weapons to accomplish the same end: eliminating competitors for the remaining food.

This will be a worldwide problem - and could easily lead to a Third World War - but Europe's vulnerability is particularly easy to analyze. The last abrupt cooling, the Younger Dryas, drastically altered Europe's climate as far east as Ukraine. Present-day Europe has more than 650 million people. It has excellent soils, and largely grows its own food. It could no longer do so if it lost the extra warming from the North Atlantic.

There is another part of the world with the same good soil, within the same latitudinal band, which we can use for a quick comparison. Canada lacks Europe's winter warmth and rainfall; it has, for example, no equivalent of the North Atlantic Current to preheat its eastbound weather systems. Canada's agriculture supports about 28 million people. If Europe had weather like Canada's, it could feed only one out of twenty-three present-day Europeans.

Any abrupt switch in climate would also disrupt food-supply routes. The only reason that two percent of our population can feed the other 98 percent is that we have a well-developed system of transportation and middlemen - but it is not very robust. The system allows for large urban populations in the best of times, but not in the case of widespread disruptions.

Natural disasters such as hurricanes and earthquakes are less troubling than abrupt coolings for two reasons: they're brief (the recovery period starts the next day) and they're local or regional (unaffected citizens can come to the assistance of the overwhelmed). There is, increasingly, international cooperation in response to catastrophe - but no country is going to be able to rely on a stored agricultural surplus for even a year, and any country will be reluctant to give away part of its surplus.

In an abrupt cooling the problem would get worse for decades, and much of the earth would be affected. A meteor strike that killed most of the population in a month would not be as serious as an abrupt cooling that eventually killed just as many. With the population crash spread out over a decade, there would be ample opportunity for civilization's institutions to be torn apart and for hatreds to build, as armies tried to grab remaining resources simply to feed the people in their own countries. The effects of an abrupt cold last for centuries. This might not be the end of *Homo sapiens* - written knowledge and elementary education might well endure - but the world after such a population crash would certainly be full of despotic governments that hated their neighbors because of recent atrocities. Recovery would be very slow.

A CHEAP-FIX SCENARIO, such as building or bombing a dam, presumes that we know enough to prevent trouble, or to nip a developing problem in the bud. But just as vaccines and antibiotics presume much knowledge about diseases, their climatic equivalents presume much knowledge about oceans, atmospheres, and past climates. Suppose we had reports that winter salt flushing was confined to certain areas, that abrupt shifts in the past were associated with localized flushing failures, and that one computer model after another suggested a solution that was likely to work even under a wide range of weather extremes. A quick fix, such as bombing ice dams, might then be possible.

Although I don't consider this scenario to be the most likely one, it is possible that solutions could turn out to be cheap and easy, and that another abrupt cooling isn't inevitable. Fatalism, in other words, might well be foolish.

A MUDDLE-THROUGH SCENARIO assumes, again, that we would mobilize our scientific and technological resources well in advance of any abrupt cooling problem, but that the solution wouldn't be simple. Instead we would try one thing after another, creating a patchwork of solutions that might hold for another few decades, allowing the search for a better stabilizing mechanism to continue.

We might, for example, anchor bargeloads of evaporation-enhancing surfactants upwind from critical downwelling sites, letting winds spread them over the ocean surface all winter, just to ensure later flushing. We might try to create a rain shadow, seeding clouds so that they drop their unsalted water well upwind of a given year's critical flushing sites.

Perhaps computer simulations will tell us that the only robust solutions are those that re-create the ocean currents of three million years ago, before the Isthmus of Panama closed off the express route for excess-salt disposal. Thus we might someday dig a wide sea-level Panama Canal in stages, carefully managing the changeover.

SCENARIOS OUGHT TO CAPTURE the stage setting that precedes the main events, not merely describe major outcomes. So let me spin a slightly more exaggerated version of our present know-something-do-nothing state of affairs: Know Nothing, Do Nothing.

My scenario doesn't require the short-sighted to be in charge, only for them to have enough influence to create starvation budgets for the relevant science agencies, to send recommendations back for yet another commission report five years hence, and so forth. In the USA, all it takes is either a know-nothing President or a do-nothing Congress.

The short-sighted have, after all, dominated at many times during history. Book burning and censorship are perhaps the most obvious signs, but countries have also withdrawn into shells, disdaining foreign inventions and knowledge in favor of fundamentalist visions of the good life (recall that Chinese retreat in 1433 from the route to Europe if you can't think of modern examples).

Most of the issues that vex humanity daily - ethnic conflict, arms escalation, overpopulation, abortion, environmental destruction, and endemic poverty, to cite several of the most persistent - can be solved only by integrating knowledge from the natural sciences with that from the social sciences and the humanities. Only fluency across the boundaries will provide a clear view of the world as it really is, not as it appears through the lens of ideology and religious dogma, or as a myopic response solely to immediate need. Yet the vast majority of our political leaders are trained primarily or exclusively in the social sciences and the humanities, and have little or no knowledge of the natural sciences. The same is true of public intellectuals, columnists, media interrogators, and think-tank gurus. The best of their analyses are careful and responsible, and sometimes correct, but the substantive base of their wisdom is fragmented and lopsided.

– EDWARD O. WILSON,
Consilience: Unity of Knowledge, 1998

[The epic learning game we call science] formalizes our special kind of collective memory, or species memory, in which each generation builds on what has been learned by those that came before, following in each other's footsteps, standing on each other's shoulders. Each generation values what it can learn from the one before, and prizes the discoveries it will pass on to the next, so that we see farther and farther, climbing an infinite mountain.

– JONATHAN WEINER, 1994

To: Human Evolution E-Seminar
From: William H. Calvin
Location: 62°N 114°W 10,400m ASL
Yellowknife, Northwest Territories

Subject: **Feedbacks in the greenhouse**

Looking down at the frozen tundra around the various chilly lakes below, I tend to see methane deposits, which isn't much better than seeing salt problems and ice dams breaking. The northern parts of Canada, alas, are exactly the areas that warm up first, in the computer simulations of global warming. A thaw from global warming is predicted to release the methane in the tundra - a greenhouse gas - and so make things still warmer yet.

Cooler can promote colder, too. Consider the effects of the high winds that blow during the cool-and-dry mode, carrying dust off the continents and fertilizing the algae in the oceans. This promotes more CO_2 removal and thus less greenhouse warming. Also, more ice at the higher latitudes makes the earth's surface whiter, and so reflects more of the sun's energy back into space, which cools things off even further. Whiter and whiter becomes colder and colder.

Indeed, as I mentioned back in Germany, the earth is now thought to have become completely covered with ice and snow on some occasions, all of the land and sea surfaces white except for a few hot springs areas such as Iceland. This may have happened multiple times back about 600-800 million years ago.

What keeps "Snowball Earth" from being permanent ("The White Earth Catastrophe") is that volcanoes still vent CO_2 to the atmosphere and that this slowly causes a greenhouse warming that finally melts back the tropical ice. There are many evolutionary implications of Snowball Earth because of the island biogeography it implies. The volcanic hot spots in the ocean also provide a wonderful route for different animals to have evolved in isolation without competing with one another for long periods, exactly what would help explain the Cambrian explosion of different body types (all those several dozen phyla) at 530 million years ago.

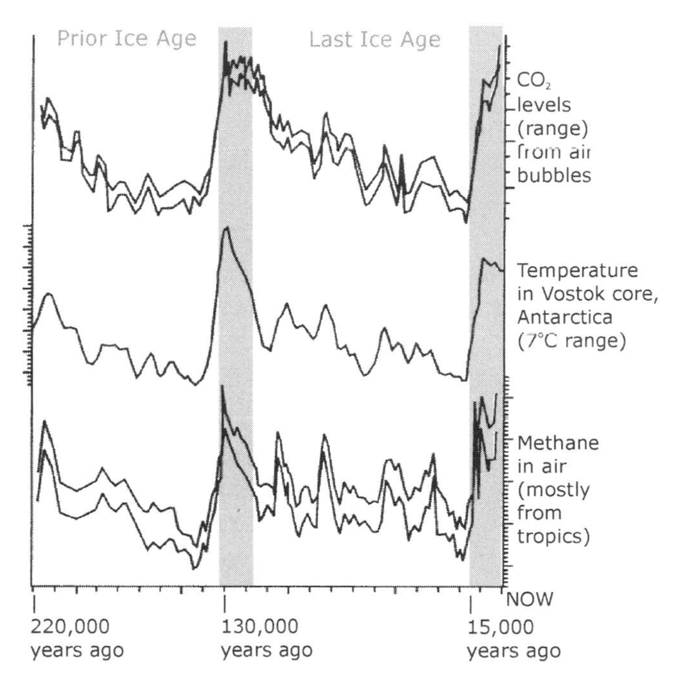

WHEREAS POSITIVE FEEDBACK BOOSTS (recall those self-perpetuating droughts back at Lake Naivasha, page 157), negative feedback fights back, serving to minimize change. If the bigger-and-bigger avalanche is the main metaphor for positive feedback, then the prosaic household thermostat is the main metaphor for negative feedback.

The classic earth sciences example of a thermostat is the way volcanic CO_2 is removed from the atmosphere by weathering rocks. The chemical reaction between the rocks and the CO_2 in the falling rain depends on temperature. The volcanic CO_2 may make things warmer via the greenhouse but that slowly increases the speed at which weathering removes the CO_2, suggesting a balancing act. Let it cool for some reason and the weathering slows; the volcanic CO_2 can build up again and so produce a greenhouse warming. However, this effect seems not to be strong enough to prevent CO_2 rising during a warm period in the ice ages.

There's an even quicker-acting possibility for negative feedback: More warming might create more clouds over the tropical oceans, their whiteness reflecting more of the sun's heat back into space and thereby cooling things somewhat. Again, no one is certain about the strength of the effect.

Some things may balance out for awhile. Indeed that is what the greenhouse effect is all about, so let me give you the short version of greenhouse physics. The earth is heated by sunlight but the warmed earth surface creates infrared radiation that escapes back out into space. The earth, however, has an atmosphere which traps some of this heat loss, as when clouds keep it from becoming so cold at night. The atmosphere is not as transparent to infrared wavelengths as it is to the incoming light (the same is true of glass, and so the same principle operates in a greenhouse or a car parked in the sun). The atmosphere's insulation blanket comes mostly from water vapor and carbon dioxide (and with minor contributions from methane and nitrous oxide and ozone), the so-called greenhouse gases.

The amount of energy coming in has to balance with what escapes (or else the earth's temperature will keep changing until it does balance). Add more greenhouse gas, and the earth will heat up until it can produce more infrared, enough so that what gets past the greenhouse gases will balance the accounts. Let the water vapor or some other greenhouse gas be reduced, and the warm earth will be losing more heat than it gains from sunlight. And so it cools. The accounts will eventually balance again at some lower temperature. There is nothing intrinsically slow about these temperature changes. You know how fast a cloudless night can cool the earth, compared to a cloudy night. Were the winds to reduce tropical evaporation by one-third in a year, you'd see an abrupt shift from warm-and-wet to cool-and-dry.

Negative feedback often overreacts and produces oscillations, which is why people are always fiddling with their thermostat. Just the right amount of negative feedback can produce homeostasis, neatly stabilizing things, but often the various competing processes operate on different time scales, making it hard to keep things balanced. Unfortunately, the climate records are full of evidence that the perfect balance is seldom achieved, certainly not on the time scale of the lifetime of a civilization.

CONCENTRATING ON ONE "CAUSE" at a time is sometimes a good strategy – and sometimes one is all the human mind can juggle, if it has lots of parts and pieces. When many causes all interact - and abrupt climate change candidates include the thermohaline circulation, the atmospheric circulation associated with the North Atlantic Oscillation, changes in tropical evaporation, and changes in albedo - the human mind needs some help.

For the big picture, you need a working model of oceans and climate. The feedbacks can be studied in the global climate models running on the big computers. To stabilize our chattering climate, we'll need to identify all of the important feedbacks that control climate and ocean currents, and estimate their relative strength and interactions.

The feedbacks are what determine the tipping point, where one mode flips into another. Near the threshold, you may sometimes see abortive responses, rather as you may step back onto the curb several times before finally running across the busy street.

Abortive responses and rapid chattering between modes are common problems in nonlinear systems with not quite enough oomph, the reason why old fluorescent lights flicker. To keep them firmly in one state or the other, you try to keep them away from the transition thresholds. That's what we need to do with climate - keep it firmly in the warm-and-wet mode, and not let it come anywhere close to the transition zone that could kick us back into the next ice age.

It's much like building flood-control dams, and whether you design for a "fifty-year flood" or a "thousand-year flood" - except that these disasters are felt around the world. For abrupt cooling, you want to make sure that no decades-to-centuries climate variations such as El Niño and the North Atlantic Oscillation can cross the threshold of the slippery slope to abrupt cooling. Of particular importance are combinations of them, which may have much larger effects than their linear sums would ordinarily suggest.

Some scientists have sought comfort by noting that extrapolating trends in atmospheric CO_2 show that the northern Gulf Stream circulation might not shut down for another 400 years. Personally I take little comfort in this, in part because my doctoral dissertation many decades ago concerned fluctuations in a threshold system (in nerve cells, but the principle applies elsewhere; in stochastic processes, it goes under the name of the "first passage time problem").

Steady trends, in real life, have fluctuations superimposed on them. Construction projects usually take longer than estimated because of delays, but quicker-than-expected is what you instead get from the climate systems. The comments about the time to shutdown of the northern Gulf Stream are about like extrapolating the slow rise in sea level to estimate when the nations of the Indian Ocean on low islands will finally be covered up. They will, of course, be rendered uninhabitable far sooner because of the fluctuations in sea level caused by unusually high tides associated with storms. Let a low pressure system be centered over the islands at an extreme high tide, and seawater will flood the islands and salt up the fresh water supply. This considerably shortens their time remaining.

The same thing holds for our present extrapolations of North Atlantic models: they simply do not yet take into account such fluctuations as El Niño and the North Atlantic Oscillation, nor freak "random" combinations of weather that set up self-perpetuating regional droughts like the Dust Bowl. None of these are likely to delay the threshold-crossing year; they are only likely to shorten the time remaining until a slippery slope is reached and the climate flips.

And besides the "natural" contributors, we have human ones associated with cutting down forests and burning fossil fuels. It's like estimating the time until the bathtub overflows when the water is left running: it makes a lot of difference if there is also a kid in the tub, bouncing around. This never delays the time until overflow; it only makes it happen sooner.

At the moment we are an ignorant species, flummoxed by the puzzles of who we are, where we came from, and what we are for. It is a gamble to bet on science for moving ahead, but it is, in my view, the only game in town.
– LEWIS THOMAS, 1979

To: Human Evolution E-Seminar
From: William H. Calvin
Location: 49.0°N 123.1°W 7,000m ASL
Bumpy border crossing

Subject: **Managing high-risk situations**

Back at the beginning of our warm period, there was mile-high ice here at the U.S.-Canadian boundary, between Victoria and Vancouver. That's because the Puget Lobe was rapidly advancing only 14,000 years ago, triggered by the great episode of rapid global warming that preceded the Younger Dryas.

When the ice sheet over the Canadian Rockies started to collapse and spread sideways, the ice came all the way down the Strait of Georgia. It went past the Strait of Juan de Fuca and on into the southern part of Puget Sound, covering Seattle's site up to the height of a 300-story building. The local mountains had their own glaciers then, as they still do, but they didn't have the major ice sheets capable of filling in between mountain ranges. It was the surge from the north that filled up the space between the Cascade Range to the east and the Olympic Mountains to the west, lasting for over a thousand years.

A few hundred miles to the east, another such ice advance blocked a river, creating a great meltwater lake (Lake Missoula) over northern Idaho and the western part of Montana. The ice dam broke, of course, with a great outburst flood surging across the middle of Washington State (we call them the "scablands" because of the deep channels and plunge pools cut in the old lava). It must have been rather loud, too. Temporarily, the Columbia River had a flow ten times greater than all the rest of the world's rivers combined.

The ice dam reformed, and broke again – at least 59 times. Just imagine the biblical flood on fast forward and stuck in a repeat loop.

We're descending now, and it's a little bumpy as we cross the San Juan Islands. Landing is always the hard part, and it's only twenty minutes from now, as I see the Strait of Juan de Fuca off to the west, stretching out into the Pacific Ocean. The end of our warm period isn't as predictable as the end of an airline flight, but it, too, surely isn't far off, if we let nature take its course. And flip again.

We humans have consciousness in a big way, compared to the great apes, enabling us to better evaluate risk, danger, and pain. We pay a big price for it, too: we can know when we can't do anything about imminent suffering.

But maybe we can do something about it.

STABILIZING OUR FLIP-FLOPPING CLIMATE is not a simple matter. We need heat in the right places, such as the Greenland Sea, and not in others right next door, such as Greenland itself. Man-made global warming is likely to achieve exactly the opposite – warming Greenland and cooling the Greenland Sea.

A remarkable amount of specious reasoning is often encountered whenever we contemplate reducing carbon-dioxide emissions. It's not a simple matter of global temperature – I hope it's now clear that no simple-minded argument in favor of *more* global warming (to "avoid cooling," or "increase agricultural productivity") should be taken seriously. These clangers are sometimes beginners' mistakes, but can also be confuse-the-issue propaganda designed to "buy time" for polluters.

That increased quantities of greenhouse gases will lead to global warming is as solid a scientific prediction as can be found, but other things influence climate too, and some people try to escape confronting the consequences of our pumping more and more greenhouse gases into the atmosphere by supposing that something will come along miraculously to counteract them. Volcanos spew sulfates, as do our own smokestacks, and these reflect some sunlight back into space, particularly over the North Atlantic and Europe. But we can't assume that anything like this will neatly counteract our longer-term flurry of carbon-dioxide emissions. Only the most naïve gamblers bet against physics, and only the most irresponsible bet with their grandchildren's resources.

To the long list of predicted consequences of global warming - stronger storms, methane release, habitat changes, ice-sheet melting, rising seas, stronger El Niños, killer heat waves - we must now add abrupt, catastrophic coolings. Whereas the familiar consequences of global warming will simply force expensive but gradual adjustments, the abrupt cooling and drying promoted by human-enhanced warming looks like a particularly efficient means of committing mass suicide via sudden agricultural collapse and the Four Horsemen that follow.

We cannot avoid trouble by merely cutting down on our present warming trend, though that's an excellent place to start. Paleoclimatic records reveal that any notion we may once have had that the climate will remain the same unless pollution changes it is, well, wishful thinking. Judging from the duration of the last warm period, we are in the declining centuries of our current one. Our goal must be to stabilize the climate in its favorable mode and ensure that enough equatorial heat continues to flow into the waters around Greenland and Norway. The stabilized climate must have a wide "comfort zone," and be able to survive the El Niños of the short term. We can design for that in computer models of climate, just as architects design earthquake-resistant sky scrapers. Implementing it might cost no more, in relative terms, than building a medieval cathedral.

Yet we may not have centuries for acquiring wisdom, and it would be wise to compress our learning into the years immediately ahead. We have to discover what has made the climate of the past 8,000 years relatively stable, and then figure out how to prop it up.

JUST BECAUSE GREAT CLIMATE FLIPS can happen in response to global warming doesn't mean that they are the most probable outcome of our current situation, what one might "forecast" (that's one of the reasons why I've been careful not to "predict" a cooling in the next century). Climate scientists have a maddening tendency to focus on "the most likely" outcome in the next century, just as economists and politicians tend to do, and I believe that is a serious mistake.

The issue here is managing high-risk situations, not the usual stuff: you can make a lot of mistakes when you try to extrapolate "business as usual." How much effort should be expended on a minority possibility, particularly one with a history of having occurred many times in the past? Other than military and disaster planners, physicians, and people in the reinsurance business, not too many people are knowledgeable about high-risk management. (I learned some only because my father was an insurance executive and I later listened to many neurosurgical conferences discuss the failure of the referring physician to think beyond "the most likely cause.")

Guessing the most likely diagnosis and outcome is something that any second-year medical student can often do. Even I can do it in some areas like neurology, just from hanging around real doctors for decades: given some symptoms and lab findings, I can sometimes say that it's most likely disease X. I might even be right half the time. But that's not good enough. The reason you consult real experts is because they know all the common mistakes and how to avoid them. And what they do, that the second-year medical student can't do yet, is to rule out the less likely scenarios, particularly the ones that could kill you quickly if left untreated. Even though such possibilities are "less likely," they're exactly what you have to focus on.

A homey example would be when you are awakened during the night to hear some strange gurgling noises. The most likely source, you realize as you lie there in bed, is simply a downspout clogged with leaves. Not a serious matter, something that can wait for a sunny weekend. But you also know that the sounds could be coming from a ruptured hose to the washing machine, and you know what a mess a flood can make, in short order. Even though there's an 80 percent chance of the noise being innocuous, you crawl out of your warm bed and go check things out.

And you see the same focus on the "less likely," both in medical diagnosis and therapy, where the physician must often act on incomplete information because of the serious consequences of delay. Suppose that with your symptoms and lab findings, the chances are 80 percent that you've got disease X, a nuisance in the long run but not catastrophic. The trouble is, the symptoms are also consistent with another more serious disease, lymphoma, which can quickly kill and needs early treatment. Even though the chances are only 20 percent that you've got lymphoma, your physician may tell you that you are going to need chemotherapy "for insurance against cancer." You can't just wait to see what develops. The possible consequences of delay are simply too great. The physician who waits until "dead certain" of a diagnosis before starting treatment may wind up with a dead patient.

I do not believe in a fate that falls on men however they act; but I do believe in a fate that falls on men unless they act.
– G. K. CHESTERTON

That's our situation with gradual warming and abrupt cooling. It isn't that abrupt cooling is the most probable outcome in the next century but that an atmospheric warming from any cause looks capable of triggering a loss of the warm water loop through the Labrador and Greenland Seas (the front-runner candidate for what has caused the observed global abrupt coolings of the past).

One should not get distracted by which-came-first issues (Is the warming "our fault"?) but focus on consequences – and particularly the possible consequences of postponing action, of simply waiting to see what develops. The failure to flush the cooled-down water from the ocean surface isn't even the 20 percent possibility at the moment; it's the best-understood candidate for what can trigger global abrupt cooling. The alternative candidates should not be used to discourage preventative action on the rapidly-fatal scenario. Promptly studying how to stabilize the North Atlantic Current ought to be high on the agenda.

Focusing on "the most likely" outcome is a beginner's mistake when the stakes are so high. Climate scientists have not, heretofore, had to cope with managing high-risk situations because they've had few interventions to offer. As that changes, thanks to the magnificent science now being done on climate, they'll need some appreciation for how to manage situations described 2,500 years ago by the Hippocratic aphorism, "Life is short, the art long, opportunity fleeting, experience treacherous, judgment difficult. The physician must be ready . . . ."

THE BOOM-AND-BUST CLIMATE CYCLE shows the challenges that our ancestors had to contend with. At the beginning of the ice ages 2.5 million years ago, our ancestors were walking upright but had ape-sized brains; they were just starting to make stone tools. By the time of our most recent ice age, we had a brain three times larger, capable of making good guesses about what's likely to happen next, and often capable of heading off trouble.

But I doubt that another episode of the bust-then-boom climate will jack us up another notch in brain power. Ever since agriculture was invented, and transportation minimized isolation, natural selection has become decoupled from human evolution, at least the ways it used to operate. We're now smart, however, in ways that owe little to our present brain power, but rather to the accumulated experience of the people that have lived since the last ice age ended. Education. Writing. Technology. Science. We accumulate knowledge and refine wisdom from it.

There's an element of "use it or lose it" here. Use our civilization's achievements to prevent the next whiplash, or lose much of civilization's gains as our warm period suddenly ends with a population crash.

Those who will not reason
Perish in the act:
Those who will not act
Perish for that reason.
– W. H. AUDEN

Afterthoughts

As I said in the postscript for my 1986 book, *The River That Flows Uphill*, "Writers sometimes feel as if they have been taken over by a book: it develops a life of its own, proclaims its own imperatives, almost writes itself once the framework is established; one has to somehow live up to its expectations."

The present framework is an amalgamation of various European and African trips, meetings, and over-the-pole flights between 1999 and 2001, rearranged to suit the thematic development. It took me several years to discover this e-seminar framework (I've never actually taught a seminar this way), as I tried to figure out how to utilize my well-edited *Atlantic Monthly* cover story, "The great climate flip-flop," in the more general context of how our ancestors evolved a chimpanzeelike brain into our present-day model. The most significant of the consequent misrepresentations is actually of me as the narrator. I certainly don't write emails which achieve the literary levels that rewriting and editors can produce.

My long interest in the brain-enhancing aspects of paleoclimate is where that 6,600-word *Atlantic Monthly* article came from, though all the brain-related materials were edited out, to focus solely upon the climate history and their oceanographic mechanisms. There's a problem with doing brain evolution too briefly, as someone (such as the climate change skeptics who so predictably enjoy being "tough-minded") will mistakenly conclude that climate catastrophes are "a good thing."

One problem I had, in writing the first draft of the *Atlantic* article in 1997, was that the facts alone were pretty depressing. These abrupt coolings were far worse than global warming, and far more frequent than meteor strikes and major volcanic coolings. But I gradually realized that I had a much more hopeful attitude towards the abrupt climate prospects – it wasn't at all like my "Why worry" attitude towards the usual potential catastrophes such as earthquakes, meteors, and Mount Rainier burying Seattle under a mud flow.

In analyzing this mismatch between the bare facts and my gut feelings, I realized that I've seen the practice of medicine change over the three decades that I've been on a medical school faculty, and it seemed to me that we ought to be able to devise interventions for climate change that are analogous to what we've done with vaccines and antibiotics – that we stand an excellent chance of being able to understand what's going on with abrupt climate change, a good chance of being able to devise strategies that "buy time," and a fair chance of developing a "vaccine" technology that can stabilize the climate so that it doesn't pop back into the cool-and-dry mode.

In the aftermath of the *Atlantic* article, when answering letters to the editor, something else fell in place. Too many people were trying to simplify things by dwelling on "the most likely scenario" for the next century. That's a classic mistake, one that physicians are trained not to get trapped by. Medical schools have institutionalized forums found in few other segments of our society, just to keep driving this home. The clinical pathological conference is where the pathologist reveals the results of the biopsy or autopsy and everyone discusses what mistakes were made in diagnosis and treatment. Thinking back, I realized that a lot of mistakes were made simply because referring physicians hadn't been able to think beyond the "most likely outcome." There, too, things have changed in recent decades and I have some hopes that a science of climate will adopt some of the ways of thinking seen in more developed areas of high-risk management.

Over the last decade, I've also gotten to know a lot of futurists, people used to thinking about the future – and about how to frame the issues. That's where the last part of this book came from, my attempts to sketch out some simple-minded interventions and my suggestions for how we can now test out such interventions on the computer models of climate change. Stabilizing the climate is one of civilization's great tasks, one that deserves much attention in the twenty-first century.

W.H.C.
Seattle

Acknowledgements

I have a lot of people to thank within the geoscience community, those who have taken the trouble to help educate an interloper from biophysics and neurobiology. The easiest to identify is the editor of *Quaternary Research*, Stephen C. Porter, because year after year he organized a lecture series at the University of Washington's Quaternary Research Center, which is where I met such people as the glaciologist Hans Oeschger in 1984 (that's where I first heard about abrupt climate change) and the archaeologist Glynn Isaac in 1983 (which is where I saw the shatter-and-search toolmaking demonstrated).

I owe particular thanks to William Whitworth, then editor of the *Atlantic Monthly*, for twisting my arm back in April, 1997. At a time when most news media were simply not reporting the abrupt climate change story, he had gotten wind of it, asked Freeman Dyson who might be able to write it, and phoned me. I spent two weeks suggesting climate researchers who could do a better job, but finally relented – mostly because I had been telling all my friends that someone ought to write this story for nonscientists, that it was a scandal that it had gone largely unreported for ten years despite all the news-feature stories in *Science* and *Nature*. Toby Lester and the other editors at the *Atlantic* did a wonderful job of improving my prose; I hope that not too many of their improvements were lost in my conversion to the present travelogue format.

Authors find it difficult to get useful feedback on first drafts and so I especially thank my volunteer readers for commenting so effectively: Ingrith Deyrup-Olsen, the late Penn Goertzel, Blanche and Seymour Graubard, Katherine Graubard, Daniel K. Hartline, Conway Leovy, India Morrison, Gordon Orians, Sonia Ragir, Susan B. Rifkin, Peter G. Rockas, and Barbara Sherer. Two of the anonymous reviewers for the University of Chicago Press were especially helpful with climate and anthropological detail.

In addition, I was aided by useful conversations with (well, it actually goes back to Melville Herskovits and Louis Leakey in 1959) Richard Alley, Elizabeth Bates, Wolf Berger, Derek Bickerton, Barry Bogin, Stewart Brand, Wally Broecker, Peter Clarke, Ronald Clarke, Meg Conkey, Iain Davidson, Terry Deacon, Derek Denton, Brian Fagan, Dean Falk, Stephen Jay Gould, William Hopkins, Richard Hutton, Glynn and Barbara Isaac, Harry Jerison, Donald Johanson, Kenneth Kidd, Richard G. Klein, Mel Konner, Kathleen Kuman, Meave Leakey, Elizabeth F. Loftus, Linda Marchant, John Maynard Smith, William McGrew, Charles Minzel, Jim Moore, Solene Morris, Toshisada Nishida, Hans Oeschger, Jay Ogilvy, David Perlmutter, Ray Pierrehumbert, Steve Pinker, Stefan Rahmstorf, Peter Rhines, Peter Richerson, Duane Rumbaugh, Ed Sarachik, Sue Savage-Rumbaugh, Margaret J. Schoeninger, Peter Schwartz, Eugenie Scott, Jeff Severinghaus, Pat Shipman, Eric Steig, Thomas Stocker, Ian Tattersall, Phillip V. Tobias, Ajit Varki, Frans de Waal, Ed Waddington, Alan Walker, Robert W. Walter, Peter Ward, Andrew Weaver, Christopher Wills, Bernard Wood, and Richard Wrangham. Several of the photographs are thanks to Qian Wang and Susan Rifkin.

I have benefited much from a book-writing stay at the Helen R. Whiteley Center at the University of Washington's Friday Harbor Laboratories and from workshops sponsored by the Mathers Foundation and the Foundation for the Future.

Glossary

The brackets [ ] indicate the page numbers where a further explanation can be found.

1,500-year cycle A minor climate cycle that varies from 1,100 to 1,800 years long, seen in rocks dropped from floating ice in the North Atlantic. It extends throughout both glacial times and the present interglaciation, though it sometimes skips a beat or two. It is difficult to see this cycle in temperature or solar variations, but its most recent cycle corresponds to the Little Ice Age between 1300 and 1865. Some occurrences may be exaggerated into a **Dansgaard-Oeschger** cycle, those climate flips from warm-and-wet to cool-and-dry. [217]

Acheulian In archaeology, a hominid toolkit (of fewer than a dozen tools) arising about 1.8 million years ago and coexisting with the older Oldowan toolkit for some time. The Acheulian handaxe (the name comes from the site in France where they were first discovered in the 19th century, St. Acheul) is the best-known exemplar.

adaptation A body-behavioral feature that has altered so that it more or less matches an environmental feature, though seldom reaching the lock-and-key extent. Curved finger bones are said to be an adaptation for hanging from tree branches while picking fruit. The loss of such arboreal adaptations occurred millions of years after the appearance of such adaptations for upright posture as indented knees and a bowl-shaped pelvis. Remember that behaviors and other functional changes can initially occur without obvious adaptations; bony changes only make some of them more efficient over the long run.

allele Alternative versions of a **gene**. Perhaps 20 percent of your expressed genes have a different allele on the other chromosome; that is, you are *heterozygous* for that gene and might switch to using it under some conditions. One reason that children are unlike parents is that parents are often passing on their less-used allele. Inbred strains have less heterozygosity.

ASL No, it's not American Sign Language but rather Above Sea Level, a map abbreviation that distinguishes elevations from Above the Surface, as in a radio tower 100 m high.

altruism Doing something for another's benefit, at expense to yourself – but not necessarily at the expense of your genes, as when you aid relatives. Beyond kin selection is reciprocal altruism, sharing with nonrelatives that is often eventually reciprocated, though the system is weakened by freeloaders ("cheaters").

bonobo *Pan paniscus,* formerly called the "pygmy chimpanzee," and the last great ape species to be identified. Until 1927, they were confused with the chimpanzee, *Pan troglodytes* - with the old common name arising because they were said to be the "chimpanzee of the pygmies," living as they do on the left bank of the Congo River in equatorial forests at about 21-22°E (they're called the "Left Bank Chimps" for other reasons, too). While bonobos are not particularly smaller, they are of a more slender body build than chimps. Humans and the other hominids last shared a common ancestor with both *Pan* species about 5-6 million years ago, what we might call "Pan prior." The two *Pan* species diverged at about the beginning of the ice ages, 2.5-3.0 million years ago, roughly the same time that the australopithecines spun off the *Homo* lineage.

bottleneck In population biology, an occasion when genetic variability in a population was greatly simplified by the loss of alternative **alleles** - not because of selection against the lost ones but simply due to reduced choice in mates. This happens when population numbers are greatly reduced for a time; the re-expansion then populates the world from the smaller range of surviving variation. [78]

cerebral cortex The outer 2 mm of the brain's cerebral hemispheres with a layered structure. It isn't required for performing a lot of simple actions but seems essential for creating new episodic memories, the fancier associations, and many new movement programs. Paleocortex, and archicortex such as hippocampus, has a simpler structure and earlier evolutionary appearance than the six-plus-layered **neocortex**.

cheater detection A feature of reciprocal **altruism** theories, what you use to identify freeloaders who receive but rarely give back.

conveyor An oceanographic term for the north-on-top, south-near-the-bottom loop of current in the Atlantic Ocean, which includes the **Gulf Stream** and the **North Atlantic Current**. After sinking, it is called the North Atlantic Deep Water on the return loop. It is also thought to extend through the Indian Ocean into the Pacific Ocean and equilibrate the large salinity differences that would otherwise develop between Atlantic and Pacific. The far-northern downwelling sites for the conveyor are in the Greenland and Labrador Seas, though **thermohaline circulation** also likely occurs at mid-northern latitudes. There is a southern upwelling site all through the southern oceans, promoted by the westerlies that circle the globe at the higher southern altitudes. [240, 247, 249, 265]

Coriolis effect The tendency (thanks to conservation of momentum) of moving objects to turn left in the Southern Hemisphere and to the right up north, and not at all at the equator. It contributes oceanic upwelling by moving the surface waters off to one side of the direction of the wind.

It also keeps the spread of heat from the tropics from proceeding directly to the poles, by turning it aside into the trade winds and creating eddies. The Northern Hemisphere low-pressure updraft eddies are counter-clockwise because the near-surface air drawn into them turns to the right. [67, 155]

creole Children invent a new language - a creole - out of the words of the pidgin **protolanguage** they hear their immigrant parents speaking. A **pidgin** is what traders, tourists, and "guest workers" (and, in the old days, slaves) use to communicate when they don't share a real language; pidgin sentences are unstructured and short, while those in creoles have the features of universal grammar.

Dansgaard-Oeschger events The abrupt warmings of 5-10°C seen throughout the last 117,000 years, superimposed on the more gradual trends of the orbital trends. The D-O cycle (abrupt warming and then abrupt cooling) is thought to occur as an exaggeration of some occurrences of the **1,500-year cycle**. The D-O's are what I have been calling the flips from the cool-dry-windy-dusty mode into warm-and-wet. [217]

DNA clock The building blocks of the genes are DNA bases, arranged in a string whose sequence carries the information about what proteins to construct. By comparing corresponding segments of DNA from two individuals, one can infer something about their degree of relatedness. Because more extensive differences require more time to happen, inferences can be made about when the two individuals last shared a common ancestor.

exaptation This was a term intended to cover cases where an organ with one original function got adapted to perform another function (such as the swim-bladders becoming lungs when the first marine creatures became terrestrial). Previously, the term "pre-adaptation" had been used, but this form is objectionable if it suggests some degree of prescience.

founder effect The great expansion from a small number of individuals discovering an empty **niche**. Like the **bottleneck** without the prior downsizing, but still having the restricted range of variation.

gene A unit of heredity, essentially that segment of a DNA molecule comprising the code for a particular peptide or protein. We also talk loosely of "a gene for blue eyes" and so forth (reification strikes again), but many a DNA gene is pleiotropic: it has multiple (and sometimes very different) effects on its body; like that maxim about intervening in complex systems, "You can't do just *one* thing."

genotype The full set of genes carried by an individual, whether expressed or silent **alleles** or junk DNA. See **phenotype** for some comparisons. What makes living matter so different from other self-organizing systems is that a cell has an information center, the genes, concerned with orchestrating the many different processes going on within the cell, and in such a manner that copies of the cell tend to survive.

GPS The Global Positioning System uses several dozen satellites in orbits about 20,000 km up. Signals from any four of them allow a GPS receiver to calculate the local latitude, longitude, and elevation above mean sea level. Differences between successive fixes allow direction of travel to be inferred, together with speed.

grammar In general use, synonymous with **syntax** but sometimes used within linguistics more narrowly, for all the three-dozen little words such as *the, with, before* that locate objects in space and time.

great apes All of the apes except the gibbon and siamang. African apes (gorilla, chimp, bonobo) are all great apes, as is the now-only-in-southeast-Asia orangutan (it too probably originally evolved in Africa).

Gulf Stream The section of the **conveyor** that loops through the Gulf of Mexico, hooks around Florida, and travels up the east coast of North America and the Grand Banks, east of which its name changes to the **North Atlantic Current**.

handaxe A misnomer in archaeological terminology, referring to a tear-shaped flattened stone tool seen starting about 1.4 million years ago and continuing to be made even 50,000 years ago. Its function is unclear; the sharpened-all-around version (which makes a good "killer frisbee") cannot be used as a handheld axe, though broken or tumbled versions might have been secondarily used for pounding and hacking functions. [145]

heat, latent The heat that is inherent in water vapor, which you get back as measurable heat when water droplets form. Similarly, when ice forms.

Heinrich events A paleoclimate term for the major ice-rafting events, where great armadas of icebergs set sail from Labrador, probably because of a sideways collapse of the Hudson's Bay ice mountain. They drop rocks from their bottoms as they melt, all while drifting southeasterly toward the Bay of Biscay. The Heinrich events are detected by the exotic rocks in layers of ocean-bottom cores in the Atlantic. They are associated with the coldest periods seen in Greenland, but sometimes the southern oceans warm up somewhat during Heinrich events, perhaps because the conveyor is totally shut down in the North Atlantic Ocean and little heat can be exported from the southern oceans.

higher intellectual functions In my terminology, a suite of *structured* functionality that encompasses **syntax**, multi-stage planning, multivoiced music, chains of logic, games with arbitrary rules, coherence-seeking (finding patterns within apparent chaos) and similar aspects of structured thought. They may all share a common neural machinery. They are all aspects of cognition difficult to discover in the **great apes** and there is some suspicion that they were largely absent before the appearance of behaviorally-modern ***Homo sapiens*** about 50,000 years ago.

hominid The paleoanthropology term for the "third chimpanzee tribe" since the common ancestor 5-6 million years ago, of which we are now the only surviving species. Hominid is used to designate the extinct australopithecine and *Homo* species, but not the chimpanzee and bonobo. *Hominoid* is the term that covers all the apes, including hominids and humans, back to the common ancestor with Old World monkeys. *Hominin* (avoided here, in agreement with Klein, 1999) is equivalent to hominid in all but purist usage.

Homo erectus The longest-lasting hominid species, first seen about 1.8 million years ago in Africa and Asia, and last seen about 50,000 years ago in China. The African branch (likely older) is now called *ergaster* rather than *erectus* by purists. *Homo* species such as *rudolfensis* go back to at least 2.4 million years ago in the East African Rift valley.

Homo heidelbergensis First seen in Ethiopia about 600,000 years ago at Bodo, it was in Europe by 500,000 years ago, sometimes in association with the earliest hearths. Its brain size, though smaller on average, was mostly within the modern range.

Homo sapiens According to DNA dating, our species arose sometime in the ice age before last, probably between 175,000 and the warm period at 125,000 years ago. "Anatomically modern" is more-or-less synonymous but is meant to contrast with "behaviorally modern" humans beginning at about 50,000 years ago in East Africa, perhaps coinciding with the most recent "out of Africa" worldwide emigration episode. The modernity transition to *Homo sapiens sapiens* may be earlier and more gradual within Africa.

innateness "Hardwired by genes" is the general idea, but it is, more generally, a bit of behavior that arises without learning. When the individual finds itself in a particular setting, out pops some complicated behavior. Mating behaviors are innate; some things are too important to be left to learning. But there isn't a dichotomy between innateness and environmental causation; as epigenetic rules show, something innate may have, via environmental triggers, profound effects on future form and function.

inheritance principle Darwin's great but often misunderstood insight, that variation is not truly random. (There would be a lot less skepticism about evolution if only people understood this.) Rather than variations being done from some ideal or average type, small undirected variations are more often done (because they survive to reproductive age) from the more successful individuals of the current generation, exploring the solution space nearby (not jumping randomly to somewhere truly unrelated) in the next generation. If the jumps were large, the starting place wouldn't matter (and the "randomness" skepticism would have some validity). [23]

island biogeography The peculiarities of animal and plant species when largely isolated, with just occasional interbreeding. An "island" can also be a deep ocean basin, a high mountain valley, or a patch in a patchy resource distribution that prevents migration because of nothing appropriate to eat on the journey. Islands often have a reduced number of species, so traditional predators or parasites may be lacking by happenstance. Species often arrive in small numbers, so **founder effects** (a special type of **bottleneck**) are a standard feature of island populations. [87]

isotopic thermometers An atom doesn't always have a standard number of neutrons in its nucleus, and so some are a little heavier than others. For example, most oxygen is an atomic weight of 16 but one "isotope" has two extra neutrons for 18 total protons and neutrons. The factory which churned out our oxygen atoms (our "local" supernova about 7 billion years ago) simply didn't standardize them. It makes little difference in ordinary chemistry (which is sensitive to the number of electrons, and thus the number of protons, but not to neutrons) but isotopes sometimes make a difference to scientific inquiries. A molecule of water with oxygen-18 is a little heavier than the usual oxygen-16 water. It doesn't evaporate as easily and, if it does, it precipitates out more easily - and when the air is colder, the effects are even more pronounced. So the ratio of oxygen-18 to oxygen-16 in an ice core, far from the site of evaporation, is something of a proxy for ancient air temperature en route. The colder it is, the fewer oxygen-18 isotopes complete the journey from seawater to glacier. The isotopic thermometer can be calibrated by borehole temperatures and by ancient snowlines. [226]

lapse rate For every 1000 meter rise in elevation, ice age air temperature dropped about 6°C. Thus, if an ancient snow line was 500 meters lower, it was likely 3°C cooler then.

Little Ice Age The minor cooling episode beginning about 1300 and lasting until about 1850, seen around the world. While a worse time for Greenland and Europe, it was a better time in East Africa. It was preceded by about eight centuries of warming, known as the Medieval warm period. They are the most recent of the events which make up the cycle which recurs every 1,100-1,800 years, whether in an ice age or interglaciation.

meme Richard Dawkins's 1976 coinage, on the analogy (with a little aid from mime and mimic) to **gene**, for a cultural copying unit, such as a new word or melody that is mimicked by others.

mitochondrial DNA Like the Y chromosome, the separate DNA of the mitochondria (the energy workhorses within cells) are not subject to shuffling during the formation of eggs and sperm, so that their changes over time are due only to point **mutations**. Most regions of the mtDNA mutate at a slow, constant rate, allowing inferences about how long it has been since two species shared a common ancestor. Your mtDNA is inherited only from your mother.

modernity See ***Homo sapiens.***

mutation An alteration in the DNA base sequence of a gene. The simplest version is a point mutation, where a single DNA base is changed.

Neandertals A robustly-built European species, *Homo neanderthalensis*, arising perhaps 400,000 years ago, occasionally pushed out of Europe by ice into Asia; in Israel they overlapped anatomically-modern *Homo sapiens* for almost 60,000 years. They largely disappeared about 35,000 years ago when behaviorally-modern ***Homo sapiens*** spread westward out of Asia into Europe. Their way of life subjected them to more bone fractures; they seldom survived until forty years of age; while making tools similar to overlapping species, there was little of the inventiveness that characterizes behaviorally-modern ***Homo sapiens***. They are named for the place where they were first found in 1856, a valley in Germany ("Neander Thal" in the German spelling of the time; the "h" has since been dropped from the German spelling of "valley" but not from the italicized species name). The valley itself was named for the seventeenth-century German poet and composer, Joachim Neander, who sought his muse there - which makes for irony, because higher intellectual functions like syntax and structured music may have been lacking in Neandertals, just as suspected for anatomically-but-not-behaviorally-modern ***Homo sapiens.***

neocortex All of **cerebral cortex** except for such places as hippocampus, the simpler layered structure that lacks the patterned recurrent excitatory connections and columnar structures that make the six-layered neocortex so interesting.

niche The "outward projection of the needs of an organism" such as food resources, migration routes, camouflage from predators, suitable housing and sites for effective reproduction. An *empty niche* is a proven niche space that is temporarily unoccupied by a tenant species.

North Atlantic Current In casual usage, most people just say **Gulf Stream**. More properly, an oceanographic term for the section of the **conveyor** beginning northeast of the Gulf Stream section, the name changing east of the Grand Banks. It continues east until turning north off Ireland and splitting into two main branches, the northern going up the Norwegian coast to sink in the Greenland Sea. The other branch turns west, goes south of Iceland, hooks around the southern tip of Greenland, and then sinks in the Labrador Sea. Once sunken, it becomes the North Atlantic Deep Water (NADW) on the southbound return loop. [241]

Out of Africa and the Multiregional Hypothesis. OOA refers to the most recent hominid spread out of Africa and the Levant in the last ice age, somewhat before 50,000 years ago. (There is also an earlier emigration out of Africa about 1.8 million years ago by ***Homo erectus***, and perhaps another at 500,000 years by ***Homo heidelbergensis***.) The genetic evidence tends to point to the 50,000-year event as being a replacement, that earlier hominids such as Neandertals (and remaining ***Homo erectus*** in east and southeast Asia) did not contribute any genes to the modern gene pool. The Multiregional Hypothesis argues that there are regional anatomical characteristics that suggest some interbreeding, while accepting that all regional varieties of modern humans are largely African. DNA evidence has contradicted many hard-evidence groupings of paleontological relationships, such as the notion several decades ago that humans were as closely related to gorillas as to chimpanzees. Whatever the importance of the Multiregional Hypothesis for modern human ancestry, similar issues cloud the interpretation of "the hard evidence" in many aspects of paleontology where there is no DNA or soft-tissue evidence available, only bones and their shapes.

Paleolithic Just as **Pleistocene** is a climate term, Paleolithic is an archaeology term for periods of stone toolmaking; they are both ways of subdividing the several million years before the present warm period featuring agriculture and civilizations. The periods of "old stone tools" in Europe are subdivided into *Lower, Middle,* and *Upper Paleolithic* periods, with the most recent ("Upper") starting perhaps 50,000 years ago when long blades appeared in Europe. Archaeologists working in Africa in the 1920s created a different system, the *Earlier, Middle,* and *Later Stone Ages*. The oldest stone tools date back to 2.6 million years ago in Ethiopia. The ESA includes the Oldowan and Acheulian tool kits; the MSA starting about 250,000 years ago encompasses flake and blade tools, made from prepared cores. The LSA, starting somewhere between 40,000 and 20,000 years ago, has microlithic technology, often with holders for the small sharp edged material. The MSA-LSA transition is often conflated with the Middle-to-Upper Paleolithic of Europe, but Africa may have made the transition earlier and more gradually. Blade technology is clearly in place in Africa by 120,000 years ago, and likely earlier.

Pan The genus of our closest living cousins among the apes, the chimpanzee (*Pan troglodytes*) and the bonobo (*Pan paniscus*). Used alone, *Pan* means both species (and perhaps our common ancestor with them). "Ancestral *Pan*" means the unnamed species between 5-6 million years ("Pan prior") and the chimp-bonobo split about 2.5-3.0 million years ago; it is analogous in use to "bipedal ape" in our branch since the common ancestor.

parcellation Fragmentation; breaking apart a population into smaller, isolated units ("parcels" or "patches"), as when rising sea level converts a hilly island into an archipelago of former hilltops. See also **island biogeography**.

phenotype Usually "body" but actually the entire constitution of an individual (anatomical, physiological, behavioral) resulting from the interaction of the genes with the environment. As Richard Dawkins emphasized in *The Extended Phenotype*, this can even include such things as the way of making bird nests or beaver lodges. Compare to **genotype**; as Peter Schuster said in his Schrödinger Lectures, "One can say that the genotype carries a program that is executed in a suitable environment and thereby produces the phenotype. The most relevant evolutionary consequence of this separation of [legislative] and executive power is that all variations through mutation and recombination occur on the genotype whereas selection operates exclusively on phenotypes."

phoneme The units of vocalization distinguished by native speakers of a language. Unlike ape calls and cries, phonemes are all meaningless by themselves, having meaning only in combinations (words). It is important to realize that phonemes are categories that standardize. For example, Japanese has a phoneme that is in between the English /L/ and /R/ in sound space. Those English phonemes are mistakenly treated by Japanese speakers as mere variants on the Japanese phoneme. Because of this "capture" by the familiar category, those Japanese speakers who can't hear the difference will also pronounce them the same, as in the familiar rice-lice confusion.

Pleistocene A climate term for the period starting about 2 million years ago, and usually subdivided into Lower, Middle, and Upper. The Middle Pleistocene starts with the reversal of the earth's magnetic field about 780,000 years ago and lasts until the beginning of the last warm period 130,000 years ago, with the Upper ending at 10,000 years ago. The Quaternary comprises the Pleistocene and the Holocene (the present warm period of the last 10,000 years). Pleistocene is generally a term for the ice ages, but their beginning keeps creeping farther back toward 3 million years ago.

Pliocene The period from 5 million years until 2 million years; the Miocene runs from 24 to 5 million years ago.

protolanguage Any form of communication that contains arbitrary, meaningful symbols but lacks syntactic structure. Forms of protolanguage include the communication of linguistically-trained apes and other animals, pidgins in their early stages, the speech of nonproficient second-language learners, and that of children under two.

pyramidal neurons The excitatory **neurons** of neocortex. They typically have a tall apical dendrite receiving input synapses and a triangular-shaped cell body (from whence the name) from which their axon leaves, creating distant output synapses. A better name would be "acacia neuron."

quasispecies The best adapted genotypes and their near neighbors. When the fitness landscape has a lot of isolating hilltops and valleys, small differences may produce reproductive isolation. The concept is best seen in virology, in how small genetic changes may allow a virus to escape the immune system's efforts to eradicate it. But it may also apply to human ancestors with high spontaneous abortion rates modulated by environmental toxins, allowing locally adapted subpopulations to remain reproductively isolated when immigrants arrive. Without such quasispeciation, immigrant genotypes would dilute out whatever local "progress" had been achieved. [120]

recombination (1) The shuffling of genetic material between an individual's two chromosome pairs that occurs just prior to the production of ova or sperm (the crossing-over phase of meiosis); and (2) the production of a new individual through the union of a sperm and an ovum from two parents at fertilization. Both the mitochrondrial DNA (mtDNA) and the Y chromosome genes escape shuffling and, because mtDNA comes only from a mother and the Y chromosome only from a father, fertilization does not change them either, making them more useful for DNA dating techniques.

refugia Regions that, in the midst of major population downsizings, still provide the essential elements of the species' niche for small subpopulations. Shorelines and mountain tops are often refugia locales, with a little bit more climate change resulting in the extinction of the subpopulation. Europe has many fewer plant species than one might expect because so many were, in effect, pushed into the Mediterranean during an ice age. [78]

Sahel In geography, it's the African transition zone between the arid Sahara and the tropical rain forests (15-20°N and 15°W-15°E). The main countries included are Senegal, Mauritania, Burkina Faso, Niger, Nigeria, and Chad. Though wet in the 1950s, the Sahel experienced an extended drought starting in the late 1960s. The Sahel droughts correlate with the sea-surface temperature cycles of the tropical Atlantic Ocean, which also determine the severity of hurricanes that reach North America.

speciation The process whereby a new species arises as a regional variant of a parent population. It usually involves a small population size. They become unable to produce fertile offspring when mating with the parent population, thus preserving whatever adaptations they had acquired in their somewhat different niche. [120]

surface-to-volume ratio It's the reason why kids have to be bundled up even when their parents don't. The rate of temperature loss is proportional to surface area, but heat capacity is proportional to volume – and babies have a lot of surface area for their volume (weight). In two dimensions, this is the perimeter-to-area ratio; for a circle, the perimeter is proportional to diameter and the area to diameter squared, so the rate of temperature change halves when the diameter doubles. Herds coming to drink at a waterhole bunch up; double the herd diameter with four times as many individuals, and while the perimeter doubles, the percentage exposed to traditional predators is cut in half. (The "killer frisbee" over-the-top strategy shows how this simple protection scheme can be turned to a predator's advantage.) [139]

Stone Age See **paleolithic**.

syntax The set of rules and principles that determine how structured sentences are formed, as when phrases and clauses can be nested inside one another. Each dialect has its own syntax, but the possibilities are not unlimited (as "universal grammar" observes).

thermohaline circulation In oceanography, the circulation path determined by temperature and salt – downwellings due to surface-water density created by low temperature and high salinity. Because dense water tends to eventually sink through less dense underlying layers, it contributes a vertical aspect to ocean currents. The sinking waters do not always mix with the underlying layers. Sometimes they slide down a continental slope to the ocean bottom (the most dense bottom waters are formed this way near Antarctica). Or they may become so dense from evaporative cooling (and evaporative augmentation of salt at the surface) that they plunge through the underlying layers. More organized thermohaline circulation occurs in giant whirlpools at some places in the Greenland Sea, 10-15 km across, slowly conveying surface waters to the ocean floor in a hard-to-see column. [234, 239]

Recommended Reading

For background reading on **climate change** in general, there are two recent books which are especially good. Neither really explores the abrupt climate flip-flops that I focus on here, but they are exactly what you might want to give policymakers to help them sort through the more general issues of climate change. Anyone who wishes to speak intelligently about ozone, greenhouse, and El Niño needs to read both of them.

GEORGE H. PHILANDER, *Is the Temperature Rising? The Uncertain Science of Global Warming* (Princeton University Press 1998). Written with grace and understatement for general readers, by someone deeply involved with modeling the climate, it covers much of a Princeton introductory course in the earth sciences.

BRIAN FAGAN, *Floods, Famines, and Emperors: El Niño and the Fate of Civilizations* (Basic Books 1999). Archaeologists have this wonderful perspective on what's gone wrong in the past, both with climate and human institutions. See also his *The Little Ice Age* (Basic Books 2000).

There are two books on a more direct lineage with this one; although neither book on anthropology and climate emphasizes the abruptness aspects, they are much better on the Miocene-Pliocene climates and the slow aspects of the Pleistocene:

STEVEN M. STANLEY, *Children of the Ice Age* (Harmony 1996). Much more on the non-abrupt aspects of anthropology and paleobiology in the ice ages.

RICK POTTS, *Humanity's Descent* (William Morrow 1996). And for an excellent review of paleoclimate indicators, see his "Environmental hypotheses of hominin evolution," *Yearbook of Physical Anthropology* 41:93-136 (1998).

There is now a new paleoclimate book, by one of the experts on the abruptness seen in the ice cores, that more directly addresses the present abruptness issues:

RICHARD B. ALLEY, *The Two-Mile Time Machine: Ice Cores, Abrupt Climate Change, and Our Future* (Princeton University Press 2000). It is written for nonspecialists (Alley has gotten a lot of practice as a frequent commentator for *Science* news articles). The book contains far more detail on abrupt climate change than the others. So if you find yourself asking, "But how could they possibly know that?" you'll find most of the answers in Alley's excellent book. Its final chapter about the future is conventional economic extrapolation, not the more relevant perspective of high-risk management seen in medicine, re-insurance, and disaster planning.

If anyone needs a quick reference about the importance with which the scientific community views the revelations about abrupt climate change, see the Perspectives in the 15 February 2000 issue of the *Proceedings of the National Academy of Sciences,* Vol. 97, No. 4, at *http://www.pnas.org/content/vol97/issue4-/index.shtml.* For a good textbook on the earth sciences, covering atmospheric sciences, geology, and oceanography, let me suggest:

BRIAN J. SKINNER, STEPHEN C. PORTER, DANIEL B. BOTKIN, *The Blue Planet: An Introduction to Earth Systems Science,* second edition (Wiley 1999).

For the **paleoanthropological side of things**, there are many choices. In general, climate makes an appearance only in its usual uplift-encouraging-the-savannas role or in the gradualist simplification of the Ice Ages. Except within the range of tree-ring dating, events that last for only a few centuries often cannot be seen in the archaeological record because of bioturbulence smoothing the record out, and the abruptness implications of the ice cores have generally not been digested yet. Besides POTTS and STANLEY, there are many other excellent books for general readers about anthropology:

JARED DIAMOND, *Guns, Germs, and Steel: The Fates of Human Societies* (W. W. Norton 1997). The focus is on the last 13,000 years and biogeography's influence on domestication.

DONALD JOHANSON and BLAKE EDGAR, *From Lucy to Language* (Simon & Schuster 1996).

RICHARD E. LEAKEY and ROGER LEWIN, *Origins Reconsidered: In Search of What Makes Us Human* (Doubleday 1992).

RICHARD E. LEAKEY, *The Origin of Humankind* (Basic Books Science Masters Series 1995).

CHRISTOPHER STRINGER and ROBIN MCKIE, *African Exodus: The Origins of Modern Humanity* (Holt 1996).

IAN TATTERSALL, *The Fossil Trail* (Oxford University Press 1995), a history of fossil finding which has excellent fossil hominid illustrations.

IAN TATTERSALL, *Becoming Human* (Harcourt Brace 1998).
IAN TATTERSALL and JEFFREY SCHWARTZ, *Extinct Humans* (Westview Press 2000).
ALAN WALKER and PAT SHIPMAN, *The Wisdom of the Bones* (Knopf 1996).

The relevant textbooks, both of which cover hominids well, are:

JOHN G. FLEAGLE, *Primate Adaptation and Evolution*, second edition (Academic Press 1999). Has more of the comparative anatomy perspective of physical anthropology.
RICHARD G. KLEIN, *The Human Career: Human Biological and Cultural Origins*, second edition (University of Chicago Press 1999). Has more of the cultural perspective of archaeology.

For aspects of the great apes, start with:

DEAN FALK, *Primate Diversity* (W. W. Norton 2000). Her book is aimed at anthropology undergraduates, with an excellent glossary.
FRANS DE WAAL, *Good Natured: The Origins of Right and Wrong* (Harvard University Press 1996). Together with his other books for general readers, such as *The Ape and the Sushi Master, Bonobo, Peacemaking Among Primates*, and *Chimpanzee Politics*, you get a good view of what the ape-human transition might have been from.
FRANS DE WAAL, editor, *Tree of Origin: What Primate Behavior Can Tell Us about Human Social Evolution* (Harvard University Press 2001). An excellent, readable collection of chapters by nine primatologists.

For the **brain side of things**, you will find many of the references in my earlier books, *The Cerebral Code, How Brains Think, Lingua ex Machina* (with the linguist Derek Bickerton), and *Conversations with Neil's Brain* (with the neurosurgeon George Ojemann), all at *http://faculty.washington.edu/wcalvin*. For the connection with behavior, see:

MELVIN KONNER, *The Tangled Wing: Biological Constraints on the Human Spirit* (W. H. Freeman 2001).

Especially for **evolutionary biology**, some fine writers have also been at work, adding to the books written by the biologists. It is, after all, one of the grand stories of all time – and nothing else makes much sense unless you understand the evolutionary process.

HELENA CRONIN, *The Ant and the Peacock* (Cambridge University Press 1992).

RICHARD DAWKINS, *Climbing Mount Improbable* (W. W. Norton 1996).

DANIEL C. DENNETT, *Darwin's Dangerous Idea* (Simon & Schuster 1995).

JONATHAN MILLER, *Darwin for Beginners* (Pantheon 1982 but much reprinted) is a fine place to get up to speed, helped by the illustrations by Borin van Loon. Called *Introducing Darwin* in some editions.

JOHN MAYNARD SMITH and EORS SZATHMARY, *The Major Transitions in Evolution* (W. H. Freeman 1995).

ERNST MAYR, *This is Biology: The Science of the Living World* (Harvard University Press 1997).

JONATHAN WEINER, *The Beak of the Finch: A Story of Evolution in Our Time* (Knopf 1994).

CHRISTOPHER WILLS, *Understanding Evolution* (W. H. Freeman 2002).

EDWARD O. WILSON, *Consilience* (Knopf 1998).

The **gradual climate change** story, about the only aspect of climate change that has reached a large audience, is not covered adequately by the present book. Few realize how strong the case is for global warming, so let me repeat here some of the items from the IPCC summary of where gradual climate change seems to be going, "Climate Change 2001: The Scientific Basis."

- The global-average surface temperature has increased over the twentieth century by about 0.6°C.
- Globally, it is very likely that the 1990s was the warmest decade and 1998 the warmest year in the instrumental record, since 1861 . . . the increase in temperature in the twentieth century is likely to have been the largest of any century during the past 1000 years.
- On average, between 1950 and 1993, night-time daily minimum air temperatures over land increased by about 0.2°C per decade. This is about twice the rate of increase in daytime daily maximum air temperatures (0.1°C per decade). This has lengthened the freeze-free season in many mid- and high-latitude regions.
- Satellite data show that there are very likely to have been decreases of about 10 percent in the extent of snow cover since the late 1960s, and ground-based observations show that there is very likely to have been a reduction of about two weeks in the annual duration of lake- and river-ice cover in the mid- and high-latitudes of the Northern Hemisphere, over the twentieth century.
- There has been a widespread retreat of mountain glaciers in non-polar regions during the twentieth century.
- Northern Hemisphere spring and summer sea-ice extent has decreased by about 10 to 15 percent since the 1950s. It is likely

that there has been about a 40 percent decline in Arctic sea-ice thickness during late summer to early autumn in recent decades and a considerably slower decline in winter sea-ice thickness.

- Tide-gauge data show that global-average sea level rose between 0.1 and 0.2 meters during the twentieth century.
- It is very likely that precipitation has increased by 0.5 to 1 percent per decade in the twentieth century over most mid- and high-latitudes of the Northern Hemisphere continents, and it is likely that rainfall has increased by 0.2 to 0.3 percent per decade over the tropical (10°N to 10°S) land areas.
- In the mid- and high-latitudes of the Northern Hemisphere over the latter half of the twentieth century, it is likely that there has been a 2 to 4 percent increase in the frequency of heavy precipitation events [thunderstorms and large-scale storm activity].
- Warm episodes of the El Niño-Southern Oscillation (ENSO) phenomenon... have been more frequent, persistent and intense since the mid 1970s, compared with the previous 100 years.
- In some regions, such as parts of Asia and Africa, the frequency and intensity of droughts have been observed to increase in recent decades.
- A few areas of the globe have not warmed in recent decades, mainly over some parts of the Southern Hemisphere oceans and parts of Antarctica.
- The atmospheric concentration of carbon dioxide has increased by 31 percent since 1750. The present CO_2 concentration has not been exceeded during the past 420,000 years and likely not during the past 20 million years. The current rate of increase is unprecedented during at least the past 20,000 years.
- About three-quarters of the anthropogenic emissions of CO_2 to the atmosphere during the past 20 years is due to fossil fuel burning. The rest is predominantly due to land-use change, especially deforestation . . . .

The entire document is at *http://www.ipcc.ch.*

Chapter Notes

When a reference is briefly given as "Author (year)," it means that the full reference will be found nearby or in the **Recommended Reading** section. These notes are available with live web links at *http://WilliamCalvin.com/BrainForAllSeasons/notes.htm.*

page

3 Alley (2000). The several-year figure is apropos of the suddenness with which the last ice age ended about 15,000 years ago in the Northern Hemisphere, e.g., "the last ice age came to an abrupt end over a period of only three years." But "big changes in less than a decade" is how paleoclimatologists typically characterize all of the abrupt warmings and coolings in the best ice cores, though the full time course from one stable state to another may take a few decades.

Even in ordinary dry spells such as the summer of 2000, "We have the hottest driest weather in perhaps 50 years, we have thousands of lightning strikes an hour, we have 300 new fires every day in the West, largely because of lightning strikes," a senior forest service official said. Reuters news story, 27 August 2000, datelined Boise, Idaho, USA.

Episodes this brief are seldom detected later by scientists studying the layers, as the worms churn the evidence, mixing up the layers from an entire millennium. Such smoothings make it impossible to tell whether a cooling developed abruptly or more slowly, and it totally hides many of the century-long abrupt coolings and droughts, which become little bumps in the record.

4 Phoenix Generation: Ovid didn't describe ashes in the Assyrian version but Hans Christian Andersen added fire in *The Phoenix Bird* (1872) version.

7 Jared Diamond, *The Third Chimpanzee* (HarperCollins 1992).

Besides the brain enlargement two million years ago, there was a further shift away from ape-like specializations; marathon-like endurance likely developed and childhood became even longer. See Leakey (1995).

11 Loren Eiseley, *The Night Country* (Scribners 1971), p.159. Alley (2000), p.83.

Darwin's home

13 Solene Morris, Louise Wilson, *Down House: The Home of Charles Darwin* (English Heritage 1998). Directions to Down House can be found at *http://WilliamCalvin.com/-bookshelf/down_hse.htm.*

An excellent biography in two volumes is Janet Browne's *Charles Darwin* (Jonathan Cape 1995, 2002). For Darwin's correspondence, see *http://www.lib.cam.ac.uk-/Departments/Darwin/.*

14 Dennett (1995), p.21.

16 "Swinging gait," see Francis Darwin, "A character sketch by Darwin's son," pp. 88-107 in his *The Life and Letters of Charles Darwin* (1887).

16 There is a nice summary of the modern synthesis and punctuated equilibria in Tattersall (1998), ch. 3.

18 Repeating catastrophes: the impact catastrophe at 65 million years ago was also a series of events and, while the extinction of the dinosaurs is one result, the adaptive radiation of mammals was another.

I don't know who used the phrase first, but George Orwell wrote "Catastrophic gradualism" in the *Common Wealth Review* (November 1945); see *Collected Essays, Journalism and Letters*, vol. 4 (Harmondsworth 1978), p.35.

20 Mayr (1997), p.189.

Evolution House, Kew Gardens

21 John Burnet, "Empedocles of Acragas" at *http://plato.evansville.edu/public/-burnet/ch5b.htm.*

23 Darwin's 1838 insight upon reading Thomas Robert Malthus, *Essay on the Principle of Population* (1798), is covered in Browne (1995) at pp.385-388.
24 A biography of Alfred Russel Wallace is at *http://www.wku.edu/~smithch/index1.htm.*
Memes are discussed in Richard Dawkins, *The Selfish Gene* (Oxford University Press, revised edition 1989); Susan Blackmore, *The Meme Machine* (Oxford University Press 1999); and William H. Calvin, "The six essentials: Minimal requirements for the Darwinian bootstrapping of quality," *Journal of Memetics* 1 (1997) at *http://faculty.washington.edu-/wcalvin/1990s/1997JMemetics.htm.*
My six essentials build on the three which Alfred Russel Wallace listed in 1875 (". . . the known laws of variation, multiplication, and heredity . . . have probably sufficed. . . ."); I make explicit the pattern, the work space competition, and the environmental biases. See Wallace's "The limits of natural selection as applied to man," chapter 10 of *Contributions to the Theory of Natural Selection* (Macmillan 1875).
25 Between a third and a half of an infant's cortical connections present at eight months of age seem to disappear by adulthood, although few neurons are lost; essentially, some axon branches are retracted. The best data is in monkey: P. Rakic, J.-P. Bourgeous, M. F. Eckenhoff, N. Zecevic, and P. Goldman-Rakic, "Concurrent overproduction of synapses in diverse regions of the primate cerebral cortex," *Science* 232:232-234 (1986). But the figure reflects the balance between creation of new synapses and breaking of old ones - and we don't yet know the rate of either creation or destruction, just some estimates of the cumulative differences. For all we know, there could be a turnover of five percent every month, with the rate of destruction only slightly greater than the rate of creation leading to the observed differences.
26 Mutations are also needed to restore variation to an inbreed population like the cheetahs who, while they might have gene-shuffling and recombination, don't have very many alternative alleles to shuffle.

Down among the fossils
27 Why ocean bottoms remain cold, see Philander (1998), p.128. A dramatic example of trapped CO_2 bubbling out in the manner of an uncapped bottle of seltzer occurs in volcanic lakes such as Lake Nyos in the mountainous region of northwestern Cameroon, where 1,700 people were killed in 1986. A strong wind causes the stratified lake to turn over; the gas-rich bottom waters, upon reaching the surface, release their gas in huge quantities. In the case of Lake Nyos in 1986, the jet of gas and water shot up about 260 feet. Moving at about 45 miles an hour, the gas reached villages 12 miles away. The lake released about a cubic kilometer of carbon dioxide; had it slowly leaked out, there would have been no problem.
It is also a sterling example of a recurring natural disaster that, now that scientists understand its mechanism, can be prevented via appropriate technology. Michel Halbwachs, Jean-Christophe Sabroux, "Removing CO_2 from Lake Nyos in Cameroon," *Science* 292(5516): 438 (20 April 2001; *http://www.sciencemag.org/cgi/content-/full/292/5516/438a*); see the 27 February 2001 *New York Times* news story, "Trying to tame the roar of deadly lakes," at *http://www.nytimes.com/2001/02/27/science/27LAKE.html.*
28 Flow through clouds, see Philander (1998), p.84. For "latent heat," see the glossary.
Tom Lehrer, see *http://www.keaveny.demon.co.uk/lehrer/lyrics*
31 Concealed ovulation, see Jared Diamond, *Why is Sex Fun?* (Basic Books 1997).
The C word: William H. Calvin, "Competing for consciousness: A Darwinian mechanism at an appropriate level of explanation," *Journal of Consciousness Studies* 5(4)389-404 (1998).
Frans de Waal, Frans Lanting, *Bonobo: The Forgotten Ape* (University of California Press 1997) . For more bonobo information, see *http://WilliamCalvin.com/teaching/bonobo.htm.*
33 Cronin (1992), p.7.

Musée de l'Homme in Paris
35 An excellent web starting point for paleoanthropology and its terminology is at *http://www.becominghuman.org*. You can find various dates for common ancestors, depending on the method and particular genes being used. Two recent references, with citations of other dating attempts, are:

F. C. Chen and W. H. Li, "Genomic divergences between humans and other hominoids and the effective population size of the common ancestor of humans and chimpanzees," *American Journal of Human Genetics* 68(2):444-456 (February 2001).

S. L. Page and Morris Goodman, "Catarrhine phylogeny: Noncoding DNA evidence for a diphyletic origin of the mangabeys and for a human-chimpanzee clade," *Molecular Phylogenetics and Evolution* 18(1):14-25 (January 2001).

Sonia Ragir, "Diet and food preparation: Rethinking early hominid behavior," *Evolutionary Anthropology* 9:153-155 (2000).

Sonia Ragir, Martin Rosenberg, Philip Tierno, "Gut morphology and the avoidance of carrion among chimpanzees, baboons, and early hominids," *Journal of Anthropological Research* 56:477-512 (2000) at *http://www.unm.edu/~jar/v56n4.html#a3*.

Mark F. Teaford and Peter S. Ungar, "Diet and the evolution of the earliest human ancestors," *Proceedings of the National Academy of Sciences (U.S.)* 97: 13506-13511 (5 December 2000).

Richard W. Wrangham, James Holland Jones, Greg Laden, David Pilbeam, and NancyLou Conklin-Brittain, "The raw and the stolen: Cooking and the ecology of human origins," *Current Anthropology* 40(5):567-594 (December 1999).

36 Frans B. M. de Waal, "Apes from Venus: Bonobos and human social evolution," in *Tree of Origin*, edited by Frans B. M. de Waal (Harvard University Press 2001), pp. 39-68.

Richard Wrangham, Dale Peterson, *Demonic Males: Apes and the Origins of Human Violence* (Houghton Mifflin 1996).

37 Kaye E. Reed, "Early hominid evolution and ecological change throughout the African Plio-Pleistocene." *Journal of Human Evolution* 32:289-322 (1997). Says australopithecines are found associated with faunas suggesting wooded habitats.

Yves Coppens, "The East Side story," *Scientific American*, pp. 88-95 (May 1994).

38 Monkeys out-competing chimps is just the latest version; there used to be lots more ape species, many more than Old World Monkeys. But the monkeys have been gaining with the Pleistocene climate changes, showing that being smarter is not always better. The Uganda story is from Michael P. Ghiglieri, *East of the Mountains of the Moon: Chimpanzee Society in the African Rain Forest* (Free Press 1988).

39 Leo Gabunia, Abesalom Vekua, David Lordkipanidze, Carl C. Swisher III, Reid Ferring, Antje Justus, Medea Nioradze, Merab Tvalchrelidze, Susan C. Antón, Gerhard Bosinski, Olaf Jöris, Marie A.de Lumley, Givi Majsuradze, Aleksander Mouskhelishvili, "Earliest Pleistocene hominid cranial remains from Dmanisi, Republic of Georgia: Taxonomy, geological setting, and age," *Science* 288:1019-1025 (12 May 2000).

C. C. Swisher, W. J. Rink, S. C. Anton, H. P. Schwarcz, G. H. Curtis, A. Suprijo and Widiasmoro, "Latest *Homo erectus* of Java: Potential contemporaneity with Homo sapiens in southeast Asia." *Science* 274: 1870-1874 (1996).

40 The scaled-down version of the multiregional hypothesis is Milford H. Wolpoff, John Hawks, David W. Frayer, Keith Hunley, "Modern human ancestry at the peripheries: A test of the replacement theory," *Science* 291:293-297 (12 January 2001).

Yuehai Ke, Bing Su, Xiufeng Song, Daru Lu, Lifeng Chen, Hongyu Li, Chunjian Qi, Sangkot Marzuki, Ranjan Deka, Peter Underhill, Chunjie Xiao, Mark Shriver, Jeff Lell, Douglas Wallace, R Spencer Wells, Mark Seielstad, Peter Oefner, Dingliang Zhu, Jianzhong Jin, Wei Huang, Ranajit Chakraborty, Zhu Chen, and Li Jin, "African origin of modern humans in east Asia: A tale of 12,000 Y chromosomes," *Science* 292:1151-1153 (10 May 2001). "We came to a simple conclusion," says Li Jin. "There are no old lineages left [from archaic Asians]."

One self-described "dedicated multiregionalist," Vince Sarich of the University of California, Berkeley, said: "I have undergone a conversion – a sort of epiphany. There are no old Y chromosome lineages [in living humans]. There are no old mtDNA lineages. Period. It was a total replacement."

41 Wisteria and Out of Africa, see Kenneth Kidd at *http://info.med.yale.edu/genetics/kkidd/wister.html*
42 H. Thieme, "Lower Paleolithic hunting spears from Germany," *Nature* 385:807-810 (1997).
43 Gordon H. Orians, "Human behavioral ecology: 140 years without Darwin is too long," *Bulletin of the Ecological Society of America* 79(1):15-28 (1998).

The English landscape architects were good at keeping the large animals in the distance. They hid a fence in a "ha-ha," a ditch through the landscape. From the customary viewpoint, the viewer didn't see the ditch, looking right over the top of it at the distant pastoral landscape.
46 Category carryover from altruism to syntax: William H. Calvin, Derek Bickerton, *Lingua ex Machina: Reconciling Darwin and Chomsky with the Human Brain* (MIT Press, 2000), chapter 10.

Bockenheim
48 Sanborn C. Brown, *Benjamin Thompson, Count Rumford* (MIT Press 1979).

Quarter-century per generation: One sometimes sees 20 years given for the human generation time. But a reasonable definition is not the shortest possible interval but the age of the mother at the birth of a child, averaged over her children that survive. With menarche at 17 in Sweden only a century ago, and with the first baby having a lower chance of survival, I'd guess that the average surviving children were mostly born when the mother was between 20 and 30.
49 Richard Dawkins, *The Selfish Gene* (Oxford University Press 1976), p.214.
52 Jeremy R. Marlow, Carina B. Lange, Gerold Wefer, Antoni Rosell-Melé, "Upwelling intensification as part of the Pliocene-Pleistocene climate transition," *Science* 290:2288-2291 (22 December 2000).
55 Leslie C. Aiello, Peter Wheeler, "The expensive tissue hypothesis. The brain and the digestive system in human and primate evolution," *Current Anthropology* 36:199-221 (1995). And see Ann Gibbon's news story, "Solving the brain's energy crisis," *Science* 380: 1345-1347 (29 May 1998).

Layover Limbo
59 "Enlarge one neocortical area, enlarge them all" paraphrased from: Barbara L. Finlay and R. B. Darlington, "Linked regularities in the development and evolution of mammalian brains," *Science* 268:1578-1584 (1995).

Variation within and between species: As my colleague Joe Felsenstein is fond of pointing out (the "coalesce fallacy"), suppose you have two species A and B. You plot brain size vs body size for a hundred As, and perhaps you get a symmetrical scatter with no trend. Ditto for B, except that Bs are usually bigger than As. If you mix up As and Bs into one big scatter plot (and don't plot the points in different colors), you get an impressive upwards trend: "bigger bodies have bigger brains," someone shouts – all without being able to see the trend within either species by itself. Maybe the trend doesn't exist at all, and is just an artifact of lumping when you should be splitting. Actually bigger bodies within a species usually do have bigger brains, but there are many situations where lumping groups can mislead you. Correlation is not causation, and sometimes correlation itself is - as with lumping the hypothetical As with the Bs - meaningless. The same caution applies, say, to plotting brain size vs. IQ scores for different geographic subpopulations, e.g., races. You need to establish the trend within the subpopulation and you constantly have to look out for a correlation which isn't cause and effect but merely a mutual consequence of some third thing such as growth rates or hormone levels at critical periods during development.
61 I earlier discussed the r-K spectrum in chapter 6 of my *The Ascent of Mind* (Bantam 1990), at *http://WilliamCalvin.com/bk5/bk5ch6.htm.*

Life history analysis: Barry Bogin, *Patterns of Human Growth, 2nd ed.* (Cambridge University Press 1999).
63 The data in the figure is adapted from figure 8.3 of Klein (1999), which is based on the 1990 collection of Aiello and Dean.

The Sahara

64 Victor Brovkin, quoted by Fred Pearce, "Violent future," *New Scientist* 2300:4-5 (21 July 2001).

65 J. Kutzbach, G. Bonan, J. Foley, S. P. Harrison, "Vegetation and soil feedbacks on the response of the African monsoon to orbital forcing in the early to middle Holocene," *Nature* 384:623-626 (19 December 1996).

J. E. Kutzbach, Z. Liu, "Response of the African Monsoon to Orbital Forcing and Ocean Feedbacks in the Middle Holocene," *Science* 278(5337) 440-443 (17 October 1997).

Martin Claussen, Claudia Kubatzki, Victor Brovkin, Andrey Ganopolski, Philipp Hoelzmann, Hans-Joachim Pachur, "Simulation of an abrupt change in Saharan vegetation in the mid-Holocene," *Geophysical Research Letters* 26(14):2037-2040 (15 July 1999).

Philipp Hoelzmann, Birgit Keding, Hubert Berke, Stefan Kröpelin and Hans-Joachim Kruse, "Environmental change and archaeology: Lake evolution and human occupation in the Eastern Sahara during the Holocene," *Palaeogeography, Palaeoclimatology, Palaeoecology* 169:193-217 (2001).

Ana Moreno, Jordi Targarona, Jorijntje Henderiks, Miquel Canals, Tim Freudenthal and Helge Meggers, "Orbital forcing of dust supply to the North Canary Basin over the last 250 kyr," *Quaternary Science Reviews* 20(12):1327-1339 (June 2001).

66 George Hadley, "Concerning the cause of the general trade winds," *Philosophical Transactions*, vol. 39 (1735).

67 For a good explanation of the Coriolis effect, see Philander (1998), pp.95-98, 234-239. This is why all ocean currents curve; the only one that seems to travel in a straight line for a long way is the Equatorial Undercurrent, 100 meters deep beneath the Pacific. Being at the equator, there is no Coriolis "force" to deflect it.

68 Wallace S. Broecker, Dorothy Peteet, Irena Hajdas, Jo Lin, Elizabeth Clark, "Antiphasing between Rainfall in Africa's Rift Valley and North America's Great Basin," *Quaternary Research* 50:12-20 (1998).

69 Stabilizing the Polar-Ferrel Cells via Gulf Stream warming at 60°N: This idea is surely not original to me, but I cannot find any serious studies of its stabilizing influence, compared to other (perhaps more important) factors.

Philander (1998), p.153.

71 Raymond Dart, "*Australopithecus africanus*: the man-ape of South Africa," *Nature* 115:195-199 (1925).

William M. Gray, John D. Sheaffer, Christopher W. Landsea, "Climate trends associated with multi-decadal variability of Atlantic hurricane activity," pp.15-53 in *Hurricanes: Climate and Socioeconomic Impacts.* (H.F. Diaz and R.S. Pulwarty, eds., Springer Verlag, New York 1997). See *http://www.aoml.noaa.gov/hrd/Landsea/climtrend/.*

72 Chimpanzees in drier areas: W. C. McGrew, P. J. Baldwin, C. E. G. Tutin, "Chimpanzees in a hot, dry and open habitat: Mt. Assirik, Senegal, West Africa," *Journal of Human Evolution* 10: 227-244 (1981); A. Kortlandt, "Marginal habitats of chimpanzees," *Journal of Human Evolution* 12: 231-278 (1983).

Hominids in arid environments, see Kaye E. Reed, "Early hominid evolution and ecological change through the African Plio-Pleistocene," *Journal of Human Evolution* 32:289-322 (1997).

Diamond (1997), p.94.

I first discussed the "pump the periphery" principle in *The Ascent of Mind* (Bantam 1990), at *http:/WilliamCalvin.com/bk5/bk5ch4.htm.*

73 Fagan (1999), pp.168-169.

74 Stanley (1996), p.109. The chimp subpopulation map is adapted from de Waal and Lanting (1997).

Latitude Zero

75 During glacial time (and perhaps during the cold flips as well), the tropics were less wet than now – but some desert regions such as Nevada were less dry.

77 Elisabeth S. Vrba, George H. Denton, Timothy C. Partridge, Lloyd H. Burckle (editors), *Paleoclimate and Evolution, with Emphasis on Human Origins* (Yale University Press 1995).

77 Stephen C. Porter, "Snowline depression in the tropics during the last glaciation," *Quaternary Science Reviews* 20(10):1067-1091 (2000). Assuming a full-glacial temperature lapse rate of -6°C/1000m, depression of mean annual temperature in glaciated alpine areas was ca 5.4±0.8°C; it is similar to values of temperature depression (5-6.4°C) for the last glaciation obtained from various terrestrial sites, but contrasts with tropical sea-surface temperature estimates that are only 1-3°C cooler than present.

Intensification of ice sheet formation: C. H. Haug, R. Tiedemann, "Effect of the formation of the Isthmus of Panama on Atlantic Ocean thermohaline circulation," *Nature* 393 (6686):673-676 (18 June 1998).

K. Billups, A. C. Ravelo, J. C. Zachos, "Early Pliocene deep water circulation in the western equatorial Atlantic: Implications for high-latitude climate change," *Paleooceanography* 13: (1) 84-95 (February 1998).

80 T. C. Johnson, C. A. Sholtz, M. R. Talbot, K. Kelts, R. D. Ricketts, G. Ngobi, K. Beuning, I. Ssemmanda, J. W. McGill, "Last Pleistocene desiccation of Lake Victoria and rapid evolution of Cichlid fishes," *Science* 273:1091-1093 (1996).

80 Tattersall (1995), pp. 219-220. Richard Potts, quoted in *Nature Science Update* (10 May 2001).

Okavango Delta

85 Frank Almeda, "Baobabs: gnarled upside-down giants," *Pacific Discovery* (Spring 1997), at *http://www.calacademy.org/calwild/pacdis/issues/spring97/wild.htm.*

Mark Twain (Samuel Langhorne Clemens), *Life on the Mississippi* (1883), at *http://docsouth.unc.edu/twainlife/menu.html*

87 Loren Eiseley, *The Night Country* (Scribners 1971), p.162.

Robert C. Walter, Richard T. Buffler, J. Henrich Bruggemann, Mireille M. M. Guillaume, Seife M. Berhe, Berhane Negassi, Yoseph Libsekal, Hai Cheng, R. Lawrence Edwards, Rudo Von Cosel, Didier Néraudeau & Mario Gagnon, "Early human occupation of the Red Sea coast of Eritrea during the last interglacial," *Nature* 405(6782)65-69 (4 May 2000). And see Chris Stringer's comments in the same issue for shorelines more generally, together with Klein (1999), p.454.

89 The aquatic ape hypothesis (I now prefer to talk of the "shoreline foraging hypothesis" to avoid the connotation of fully aquatic) and the associated physiological agenda were reviewed at Mile 136 of my *The River That Flows Uphill* (Macmillan 1986), available at *http://WilliamCalvin.com/bk3/bk3day9.htm.* Physical anthropologists are unbelievably vehement in their rejection of any aquatic aspect; they talk of only "fringe groups" believing it with about the tone they otherwise reserve for those audience questions about humans evolving from space travelers, so it is not surprising that their research students stay away from the subject.

Phillip V. Tobias, "Water and human evolution," *Out There* 35: 38-44 (1998) at *http://archive.outthere.co.za/98/dec98/disp1dec.html* analyzes the savannah hypothesis and why it proved flawed, emphasizing the unanswered questions about water.

For a neurobiological agenda that is also usually missing, see the brain developmental trajectories in Terry Deacon's *The Symbolic Species* (W. W. Norton 1997).

90 For an excellent example of how predation changes the rates of somatic growth and reproductive maturity (and makes bodies much larger and longer-lasting), see Todd A. Crowl and Alan P. Covich, "Predator-induced life-history shifts in a freshwater snail," *Science* 247:949-951 (23 February 1990). For a survey of dwarf species on the Mediterranean islands, see Paul Sondaar, "The island sweepstakes," *Natural History* 95(9):50-57 (September 1986).

During the last interglacial about 128,000 years ago, a range of hills in western Normandy was isolated by rising sea level, becoming the island of Jersey. And within a time span of only 6,000 years (during which mainland deer didn't change – and hadn't for the previous 400,000 years, either), the body size of the deer inhabiting the island dropped to about one-sixth of their original size: A. M. Lister, "Rapid dwarfing of red deer on Jersey in the last interglacial," *Nature* 342:539-542 (30 November 1989).

93 Desmond Tutu, quoted in "A Tutu tribute," *South African Airways* magazine (November 2000), p.30.

93 William D. Hamilton, "The genetical evolution of social behavior," *Journal of Theoretical Biology* 7:1-52 (1964).

Robert Trivers, "The evolution of reciprocal altruism," *Quarterly Review of Biology* 46:35-57 (1971).

Elliott Sober, David Sloan Wilson, *Unto Others: The Evolution and Psychology of Unselfish Behavior* (Harvard University Press 1998).

Peter J. Richerson & Robert Boyd, "The Pleistocene and the origins of human culture: Built for speed," In *Perspectives in Ethology,* Volume 13. Nicholas S. Thompson and Francois Tonneau, eds. (Kluwer Academic/Plenum Publishers, New York. 2000), pp. 1-45.

94 Glynn Isaac, as quoted by Leakey & Lewin (1992), p. 181.

Sossusvlei Dunes

98 The fauna associated with Australopithecine fossils indicate a wooded environment (Reed 1997, p.318); their *Paranthropus* successors were sometimes found in wetland environments, but it is only the later *Homo* species (*ergaster, erectus*) that are found in extremely arid and open landscape.

99 Marlow, et al (2000). The upwelling is from a depth of about 200 m, forming filaments of cold nutrient-rich waters that extend well offshore and mix with low-productivity oceanic water, forming a zone of year-round high phytoplankton productivity.

The figure is adapted from N. J. Shackleton, "New data on the evolution of Pliocene climatic variability," in E. S. Vrba, G. H. Denton, T. C. Partridge, and L. H. Burckle (eds.), *Paleoclimate and Evolution, With Emphasis on Human Origins* (Yale University Press 1995), pp. 242-248.

100 Walker and Shipman (1996), pp.89-93.

Sterkfontein Caves

103 K. Kuman and R. J. Clarke, *Sterkfontein Caves: A summary of scientific research,* a pamphlet available at *http://cyberfair.gsn.org/adelaar/sterkfon.htm.* Sterkfontein is also nicely described by Walker and Shipman (1996), p.95. For the World Heritage Site, see *http://www.cradleofhumankind.co.za.*

The Taung endocast showing the brain surface is a natural one, made by concretelike sediments collecting on the inside of the skull long after death and settling into whatever grooves the brain surface and blood vessels had imprinted on the inside of the skull. Latex endocasts can also be made of skulls. For a photo gallery of skulls and endocasts, see Dean Falk's collection at *http://www.albany.edu/braindance/gallery.htm.* A comparative collection of brains is at *http://brainmuseum.org.* The photograph of Tobias and Calvin is thanks to Dr. Qian Wang.

104 Ronald J. Clarke, Phillip V. Tobias, "Sterkfontein Member 2 foot bones of the oldest South African hominid," *Science* 269:521-524 (28 July 1995). And see *http://cyberfair.gsn.org/adelaar/little.htm.*

Ronald J. Clarke, "First ever discovery of a well-preserved skull and associated skeleton of *Australopithecus,*" *South African Journal of Science* 94:460-464 (October 1998) at *http://www.nrf.ac.za/sajs/sp_oct98.stm*

Ronald J. Clarke, "Discovery of the complete arm and hand of the 3.3 million-year-old *Australopithecus* skeleton from Sterkfontein," *South African Journal of Science* 95:477-480 (November/December 1999) at *http://www.nrf.ac.za/sajs/sp_nov99.stm*

106 Growth curves in culture, see Christophe Boesch, Michael Tomasello, "Chimpanzee and human cultures," *Current Anthropology* 39:591-614 (December 1998). They introduce an unfortunate terminology for growth curve, calling it the "ratchet effect" when what they mean has no necessary aspect of backsliding prevention, only of accretion and elaboration.

William C. McGrew, "The nature of culture: Prospects and pitfalls of cultural primatology," *Tree of Origin,* edited by Frans B. M. de Waal (Harvard University Press 2001), pp.231-254.

Henrique Teotónio, Michael R. Rose, "Variation in the reversibility of evolution," *Nature* 408:463-466 (23 November 2000).

Tasmanian cultural losses: see Diamond (1997), pp.312-313.

107 Richard W. Wrangham, "Out of the *Pan*, into the fire: How our ancestors' evolution depended on what they ate," in *Tree of Origin*, edited by Frans B. M. de Waal (Harvard University Press 2001), pp.121-143.

Craig B. Stanford, "The ape's gift: Meat-eating, meat-sharing, and human evolution," in *Tree of Origin*, edited by Frans B. M. de Waal (Harvard University Press 2001), pp.97-117.

110 Alfred Russel Wallace, *The Malay Archipelago: The Land of the Orang-utan and the Bird of Paradise; A Narrative of Travel With Studies of Man and Nature* (Macmillan 1869).

Cape of Good Hope

111 South African Museum web page (*http://www.museums.org.za/sam/resource/arch/past.htm*): "Tidal Fish Traps: fish traps are artificial tidal pools constructed of boulders across gullies in the intertidal zone of rocky shores. In the recent past some were rebuilt and used by local landowners. A number were destroyed during the construction of tidal swimming pools. They may date back some 1600 to 2000 years when the first pastoralists reached the western Cape."

112 Coastal caves, see Klein (1999), p. 456.

113 Louise Levathes, *When China Ruled the Seas: The Treasure Fleet of the Dragon Throne, 1405-1433* (Oxford University Press 1994). For more, see *http://www.cronab.demon.co.uk/china.htm*.

114 Jared Diamond, *Guns, Germs and Steel: Fates of Human Societies* (W. W. Norton 1997).

Nairobi

117 See Tattersall (1995), p. 231, for a rationale for even more hominid species and genera.

An excellent starting-point for savanna reading is Peter Matthiessen, *The Tree Where Man Was Born* (E. P. Dutton 1972).

119 Humans lost their "breeding season" some time ago with the advent of "concealed" ovulation (the loss of estrus behaviors which advertise optimal fertility), but similar "analog" changes are likely in many other areas capable of yielding reproductive isolation.

Tattersall (1998), p.103.

120 For a discussion of imprinting and its role in speciation, see P.B. Vrana, J.A. Fossella, P. Matteson, T. del Rio, M.J.O. O'Neill, S.M. Tilgham, "Genetic and epigenetic incompatibilities underlie hybrid dysgenesis in *Peromyscus*," *Nature Genetics* 25:120-125 (2000).

Various studies estimate the number of unsuspected pregnancies which abort early at 13-31 percent, though in some groups (such as women who have had pelvic inflammatory disease), "early pregnancy loss" can rise to 70 percent. So far, it seems to be the conceptions with normal karyotype whose abortion rate is increased; this isn't just the obviously abnormal ones (trisomies, etc.).

Five cups of coffee per day doubles the spontaneous abortion rate; see Sven Cnattingius et al., "Caffeine intake and the risk of first-trimester spontaneous abortion," *New England Journal of Medicine* 343(25):1839-1845 (December 21, 2000). Five-fold increases in spontaneous abortion are seen in some California counties if women drink the tap water rather than bottled water; see S. H. Swan et al., "A prospective study of spontaneous abortion: relation to amount and source of drinking water consumed in early pregnancy," *Epidemiology* 9(2):126-133 (March 1998). There are now a number of studies about tobacco and alcohol increasing spontaneous abortions but one must be careful to distinguish studies whose patient population has a confirmed pregnancy (seven weeks after last menses) from those "early pregnancy loss" studies using daily urine samples tested for human chorionic gonadotropin (hCG) to detect pregnancy via hCG rise in the second week after ovulation. It is the abortion rate in the first six weeks which is so high and unexpected, a rate comparable to the induced abortion rate in many societies.

121 For the hyrax cooperation story, see Walker and Shipman (1996), pp.151-152.

122 The D2 dopamine allele story is Kenneth Blum, John G. Cull, Eric R. Braverman and David E. Comings, "Reward deficiency syndrome," *American Scientist* 84(2): 132ff (March-April 1996) at *http://www.sigmaxi.org/amsci/Articles/96Articles/Blum-full.html*.

123 Richard Dawkins, *Climbing Mount Improbable* (W. W. Norton, 1996), p.326.

Olorgesailie

125 A readable introduction to hominid tool use is Stanley H. Ambrose, "Paleolithic technology and human evolution," *Science* 291(5509)1748-1753 (2 March 2001).

126 Glynn L. Isaac, assisted by Barbara Isaac, *Olorgesailie: Archaeological Studies of a Middle Pleistocene Lake Basin in Kenya* (University of Chicago Press 1977). And see *The Archaeology of Human Origins: Papers by Glynn Isaac*, Barbara Isaac (editor), pp. 289-311 (Cambridge University Press 1989).

Richard Potts, Anna K. Behrensmeyer, Peter Ditchfield, "Paleolandscape variation and Early Pleistocene hominid activities: Members 1 and 7, Olorgesailie Formation, Kenya," *Journal of Human Evolution* 37:747-788 (1999).

128 When I first saw a video of this torsional fracture technique for extracting the whitish-pink bone marrow from a fresh kill, I finally understood why the Latin prefix *myelo-* was used for both bone marrow and for the spinal cord. They look alike: long, pink tubes. Well, the fresh spinal cord has a lot of little cut nerves, once you've extracted it, but the resemblance to fresh bone marrow is quite striking.

129 Scavenging of contaminated meat (rather than the bone marrow and amputated legs of my examples) is perhaps best left to animals with specialized guts; see Ragir et al (2000).

130 The handaxe drawings by C. O. Waterhouse are adapted from p.70 of Kenneth P. Oakley's *Man the Tool-maker* (University of Chicago Press 1949). The somewhat rounder Acheulian handaxe is from St. Acheul, near Amiens in the Somme. The more elongate Acheulian handaxe is from Olorgesailie.

Apropos "factory site" interpretations: The density of handaxes is not what they would have looked like in any one year, long ago. The hard surface on which they rest was what didn't erode, when rains washed away the mud from layers above. So handaxes from many now-missing layers dropped their handaxes down onto the surfaces which survived the erosion (losing, of course, any vertical orientation they might have had from landing on edge). Still, that's a lot of lost handaxes - and this deflational surface is likely what was seen by visiting *Homo erectus* at various times: an impressive display of *objets trouvé*.

131 Potts (1996), p.144.

Kariandusi

133 The Rift Valley diagram is modeled after one in the National Museums of Kenya.

136 *Science* cover photo by R. Potts and W. Huang, showing opposing sides of a bifacially flaked large cutting tool (~803,000 years old) from the Bose Basin, South China: Hou Yamei, Richard Potts, Yuan Baoyin, Guo Zhengtang, Alan Deino, Wang Wei, Jennifer Clark, Xie Guangmao, Huang Weiwen, "Mid-Pleistocene Acheulean-like stone technology of the Bose Basin, South China," *Science* 287(5458):1622-1626 (3 March 2000).

Barbara Isaac, "Throwing and human evolution," *The African Archaeological Review* 5:3-17 (1987).

Potts (1996), p.139.

137 Thomas G. Wynn, "Handaxe enigmas," *World Archaeology* 27:10-23 (1995).

Kathy D. Schick, Nicholas Toth, *Making Silent Stones Speak: Human Evolution and the Dawn of Technology* (Simon & Schuster 1993).

138 Eileen M. O'Brien, "The projectile capabilities of an Acheulian handaxe from Olorgesailie," *Current Anthropology* 22:76-79 (1981). And "What was the Acheulean hand ax?" *Natural History* 93:20-23 (1984). The original idea about the handaxe being thrown may be M. D. W. Jeffreys, "The hand bolt," *Man* 65:154 (1965).

Handaxes in dried-up ponds and watercourses: Isaac (1977); F. Clark Howell, "Isimila: A Paleolithic site in Africa," *Scientific American* 205:118-129 (1961); M. R. Kleindienst and C. M. Keller, "Towards a functional analysis of handaxes and cleavers: The evidence from East Africa," *Man* 11:176-187 (1976).

Hunting per se: Matt Cartmill, *A View to a Death in the Morning* (Harvard University Press 1993).

138 William H. Calvin, "The unitary hypothesis: A common neural circuitry for novel manipulations, language, plan-ahead, and throwing?" pp. 230-250 in *Tools, Language, and Cognition in Human Evolution*, edited by Kathleen R. Gibson and Tim Ingold (Cambridge University Press 1993) at *http://faculty.washington.edu/wcalvin/1990s/1993Unitary.htm.*
140 There is one report of chimps throwing with a predatory result, see Frans X. Plooij, "Tool-use during chimpanzee's bushpig hunt," *Carnivore* 1:103-106 (1978).
143 William H. Calvin, "Rediscovery and the cognitive aspects of toolmaking: Lessons from the handaxe," short commentary at *http://faculty.washington.edu/wcalvin-/2001/handaxe.htm*

Lake Nakuru
147 Good introductions to climate and human evolution include Potts (1996); Stanley (1996); Reed (1997); and Stanley H. Ambrose, "Late Pleistocene human population bottlenecks, volcanic winter, and differentiation of modern humans," *Journal of Human Evolution* 34(6):623-651 (1998). For Africa more generally, see John Reader, *Africa: A biography of a continent* (Knopf 1998).

Perhaps that's why I like acacias so much, having studied cortical neurons for so long. Besides the branching pattern similarities, they both have thorns ("dendritic spines"). Indeed, "pyramidal neuron" is another misnomer (they only look triangular if their branching dendrites are invisible, as they usually were a century ago with the early microscopic techniques).

Lake Baringo
151 Andrew Hill, "Faunal and environmental change in the neogene of East Africa: Evidence from the Tugen Hill Sequence, Baringo District, Kenya," in Vrba (1995), pp.178-193.

Brigitte Senut, Martin Pickford, D. Gommery, P. Mein, C. Cheboi, and Yves Coppens, "First hominid from the Miocene (Lukeino formation, Kenya)," *C. R. Acad. Sci., Paris* 332, 137-144 (2001), at *http://www.becominghuman.org/news_views/article_assets/CRAS-1.PDF.*

Yohannes Haile-Selassie , "Late Miocene hominids from the Middle Awash, Ethiopia," *Nature* 412:178-181 (12 July 2001).

Lake Naivasha
156 Stanley H. Ambrose, "Chronology of the Later Stone Age and Food Production in East Africa ," *Journal of Archaeological Research* 25(4):377-392 (1 April 1998). "Enkapune Ya Muto rockshelter... contains the oldest known archaeological horizons spanning this transition [to the Later Stone Age]. Radiocarbon and obsidian hydration dates from this 5-6 m deep cultural sequence show that the Later Stone Age began substantially earlier than 46,000 years ago."
157 Dirk Verschuren, Kathleen R. Laird, Brian F. Cumming, "Rainfall and drought in equatorial East Africa during the past 1,100 years," *Nature* 403:410-414 (27 January 2000).

"Bad times," see J. B. Webster, in *Chronology, Migration and Drought in Interlacustrine Africa* (J. B. Webster, editor), pp. 1-37 (Longman & Dalhousie University Press 1979).

Gerard Bond, W. Showers, M. Cheseby, R. Lotti, P. Almasi, P. deMenocal, P. Priore, H. Cullen, I. Hajdas, G. Bonani, "A pervasive millennial-scale cycle in North Atlantic Holocene and glacial cycles," *Science* 278:1257-1266 (1997).
158 Droughts: Peter B. deMenocal, "Cultural responses to climate change during the late Holocene," *Science* 292:667-673 (27 April 2001). U. S. map at *http://www.sciencemag.org/cgi/content/full/292/5517/667/F1.*

The story of the Dust Bowl drought is at *http://www.ngdc.noaa.gov/paleo/drought/drght_home.html*

John Steinbeck, *The Grapes of Wrath* (Viking 1939), p.2.

Fidel A. Roig, Carlos Le-Quesne, José A. Boninsegna, Keith R. Briffa, Antonio Lara, Håkan Grudd, Philip D. Jones & Carolina Villagrán, "Climate variability 50,000 years ago in mid-latitude Chile as reconstructed from tree rings," *Nature* 410(6828): 567-570 (29 March 2001), at *http://www.nature.com/nlink/v410/n6828/abs/410567a0_fs.html.*

159 The self-perpetuating drought treatment is adapted from a manuscript by my colleague John Michael Wallace, "Role of the atmosphere in abrupt climate change," at *http://www.geophys.washington.edu.*

Jeffrey P. Severinghaus and Edward J. Brook, "An overview of the precise timing of atmospheric methane change relative to abrupt Greenland climate events in the past 60 ky," American Geophysical Union fall meeting (2000).

160 Climate pumping along Silk Road route to China: Adam Chou, "Migration of early hominids during the Pleistocene," *Journal of Human Evolution* 40(3):A5 (March 2001).

161 Greenland colony: John Gribbin, Mary Gribbin, *Children of the Ice* (Basil Blackwell 1980).

For group selection, see Sober & Wilson (1998), the special issue of *American Naturalist* edited by David Sloan Wilson, or his short piece, "Human groups as units of selection," *Science* 276:1816-1817 (20 June 1997) at *http://www.sciencemag.org-/cgi/content/full/276/5320/1816.*

163 Finlay & Darlington (1995).

William H. Calvin, "The emergence of intelligence," *Scientific American* 271(4):100-107 (October 1994; also appears in *Life in the Universe,* 1995), at *http://WilliamCalvin.com /1990s/1994SciAmer.htm.*

William H. Calvin, "A stone's throw and its launch window: Timing precision and its implications for language and hominid brains," *Journal of Theoretical Biology* 104:121-135 (1983) at *http://WilliamCalvin.com/1980s/1983JTheoretBiol.htm.* The modern version of my throwing theory is in Calvin (1993).

William H. Calvin and Derek Bickerton, *Lingua ex Machina: Reconciling Darwin and Chomsky with the Human Brain* (MIT Press 2000), at *http://WilliamCalvin.com/LEM.*

165 Nicholas Humphrey's book *The Inner Eye* (Faber and Faber 1986) is a good exposition on the role of social life in shaping up intelligence.

168 Frans de Waal, *Good Natured: The Origins of Right and Wrong* (Harvard University Press 1996), p.146.

Olduvai Gorge

171 James C. Woodburn, "An introduction to Hadza ecology," in *Man the Hunter,* edited by Richard B. Lee and Irven DeVore (Aldine 1968) , pp. 19-55.

172 Peter Matthiessen, *The Tree Where Man Was Born* (E. P. Dutton 1972), pp. 285-287.

173 Science literacy, see the National Science Foundation's report, "Science & Engineering Indicators 1998," chapter 7, at *http://www.nsf.gov/sbe/srs/seind98/pdf/c7.pdf.*

Theodosius Dobzhansky outlined the importance of evolution in the teaching of biology in an issue of *The American Biology Teacher*: "Seen in the light of evolution, biology is, perhaps, intellectually the most satisfying and inspiring science. Without that light it becomes a pile of sundry facts, some of them interesting or curious but making no meaningful picture as a whole."

Maasai Mara

178 Phillip V. Tobias, "Water and human evolution," *Out There* 35: 38-44 (1998) at *http://archive.outthere.co.za/98/dec98/disp1dec.html*

Splashing, see Richard J. Parnell, Hannah M. Buchanan-Smith, "Animal behaviour: An unusual social display by gorillas," *Nature* 412:294 (19 July 2001). Large numbers of gorillas feed in open, swampy clearings in the forest of northern Congo.

Thure E. Cerling, "Development of grasslands and savannahs in East Africa during the Neogene," *Paleogeography, Paleoclimatology, Paleoecology* 97:241-247 (1992).

179 M.J. Wooller, F.A. Street-Perrott, A.D.Q. Agnew, "Late Quaternary fires and grassland palaeoecology of Mount Kenya, East Africa: evidence from charred grass cuticles in lake sediments," *Palaeogeography, Palaeoclimatology, Palaeoecology* 164:223-246 (December 2000).

Documenting ancient fires is, like finding evidence for human use of fire, more difficult than it first appears. Winds blow, worms churn, habitation sites erode from foot traffic - all serve to mix annual layers in any core except an ice or tree core. When the diffusion process is regular, in the manner of heat within an ice sheet, deconvolution techniques can sometimes recover the original depth profile, as has been done for borehole temperatures in ice. But even if protected pockets of sediment could be found with the potential for high temporal resolution, the issue is really such evidence on a continental scale. Habitats are always being asynchronously disrupted by fire. The issue for present purposes is how extensive and simultaneous a fire episode is, for which we must presently rely on the widespread and quasi-synchronous warm-and-wet and cool-and-dry associations.

181 Packrats, from Donald F. Hoffmeister, *Mammals of the Grand Canyon* (University of Illinois Press 1971).

184 Patricia K. Kuhl, S. Kirtani, T. Deguchi, A. Hayashi, E. B. Stevens, C. D. Dugger, and P. Iverson, "Effects of language experience on speech perception: American and Japanese infants' perception of /ra/ and /la/," *Journal of the Acoustical Society of America* 102:3135 (1997).

Tattersall (1998), p.102.

186 Jordi Sabater-Pi, J. J. Véa, J. Serrallonga, "Did the first hominids build nests?" *Current Anthropology* 38(5):914-916 (1997).

188 Amazon flow in Younger Dryas: Mark A. Maslin and Stephen J. Burns, "Reconstruction of the Amazon Basin effective moisture availability over the past 14,000 years," *Science* 290: 2285-2287 (22 December 2000).

Libya by moonlight

195 Sally McBrearty and Alison S. Brooks, "The revolution that wasn't: A new interpretation of the origin of modern human behavior," *Journal of Human Evolution* 39 (5):453-563 (November 2000), at p.492.

Leakey & Lewin (1992), p.212.

196 Derek Bickerton, *Language and Species* (University of Chicago Press 1990).

Leakey (1995), p.93, and Michael Balter, "New light on the oldest art," *Science* news article, 283(5404): 920-922 (12 February 1999). The radiocarbon date for Chauvet is 31,000 years; the calibrated dates range from 33,000 to 38,000 years. See Edouard Bard, "Extending the calibrated radiocarbon record," *Science* 292:2443-2444 (29 June 2001).

197 Ian Tattersall, "The origin of the human capacity," *68th James Arthur Lecture on the Evolution of the Human Brain*, American Museum of Natural History (1998), at *http://www.amnh.org/enews/head/e1_h15b.html.*

198 Automaticity, see John Bargh's work at *http://www.psych.nyu.edu/people/-faculty/bargh/.*

McBrearty & Brooks (2000).

199 Ganglike attacks by male chimpanzees on isolated neighbors, see Wrangham & Peterson (1996).

200 "These results indicate that male movement out of Africa first occurred around 47,000 years ago. The age of mutation 2, at around 40,000 years ago, represents an estimate of the time of the beginning of global expansion." Russell Thomson, Jonathan K. Pritchard, Peidong Shen, Peter J. Oefner, and Marcus W. Feldman, "Recent common ancestry of human Y chromosomes: Evidence from DNA sequence data," *Proc. Natl. Acad. Sci. USA* 97(13):7360-7365 (20 June 2000).

Max Ingman, Henrik Kaessmann, Svante Pääbo, Ulf Gyllensten, "Mitochondrial genome variation and the origin of modern humans." *Nature* 408:708-713 (7 December 2000).

Ornella Semino, Giuseppe Passarino, Peter J. Oefner, Alice A. Lin, Svetlana Arbuzova, Lars E. Beckman, Giovanna De Benedictis, Paolo Francalacci, Anastasia Kouvatsi, Svetlana Limborska, Mladen Marcikia, Anna Mika, Barbara Mika, Dragan Primorac, A. Silvana Santachiara-Benerecetti, L. Luca Cavalli-Sforza, and Peter A. Underhill, "The genetic legacy of paleolithic *Homo sapiens sapiens* in extant Europeans: a Y chromosome perspective," *Science* 290:1155-1159 (10 November 2000).

David Pilbeam, "Hominoid systematics: The soft evidence," *Proceedings of the National Academy of Sciences* 97(20):10684-10686 (26 September 2000).

200 Early dates: remember that recombination can severely skew the binary-branching trees we construct. Were there some horizontal gene transfer affecting the Y chromosome or the mtDNA, it would make groups that didn't participate look like they were the root of the tree.
201 Sahara suddenness, see references on page 321.
204 Larry D. Agenbroad, "New World mammoth distribution," chapter 3 of Paul S. Martin, Richard G. Klein, editors, *Quaternary Extinctions* (University of Arizona Press 1984), pp. 90-108.

Tom D. Dillehay, "The late Pleistocene cultures of South America," *Evolutionary Anthropology* 7:206-216 (1999).

African skull shapes at 10,000 years ago in South America: Walter A. Neves, M. Blum, A. Prous, J. Powell, "Paleoindian skeletal remains from Santana do Riacho I, Minas Gerais, Brazil: Archaeological background, chronological context and comparative cranial morphology," *American Journal of Physical Anthropology* 114(S32):112-113 (2001).

Layover Limbo (again)
205 Fagan (2000), p.214. Another readable account of the Little Ice Age is Thomas Levenson, *Ice Times* (Harper & Row 1989), chapter 4.
208 Puritans, see Jacques Barzun, *From Dawn to Decadence* (HarperCollins 2000), pp.277-283.
209 Little Ice Age and witches, see Fagan (2000), p.91.

Johann Wolfgang von Goethe, *Maxims and Reflections I.*

Carl Sagan, *The Demon-haunted World* (Random House 1996), p.26.

210 My sketch of the Gulf Stream and the underlying return current (the North Atlantic Deep Water) is adapted from that of Stefan Rahmstorf, "Risk of sea-change in the Atlantic," *Nature* 388:825-826 (28 August 1997).

Copenhagen's ice cores
212 There are two fine popular treatments of climate change from the 1980s that contain much useful background for today's general reader:

John Imbrie and Katherine P. Imbrie, *Ice Ages* (Harvard University Press 1986).

Thomas Levenson, *Ice Times: Climate, Science, and Life on Earth* (Harper & Row 1989).

The beginning of the ice age at 2.5 million years is dated by N. J. Shackleton, J. Backman, H. Zimmerman, D. V. Kent, M. A. Hall, D. G. Roberts, D. Schnitker, J. G. Baldauf, A. Desprairies, R. Homrighausen, P. Huddlestun, J. B. Keene, A. J. Kaltenback, K. A. O. Krumsiek, A. C. Morton, J. W. Murray, and J. Westberg-Smith, "Oxygen isotope calibration of the onset of ice-rafting and history of glaciation in the North Atlantic region." *Nature* 307:620-623 (1984). But, as would be expected from their origins in the earth's orbital cycles, the Milankovitch rhythms were present long before that, and can be seen as cycles of deep-sea anoxia: T. D. Herbert and A. G. Fischer, "Milankovitch climatic origin of mid-Cretaceous black shale rhythms in central Italy." *Nature* 321:739-743 (1986). The precession and tilt (though not eccentricity) rhythms were faster, back when the moon's orbit was closer to earth: Andre Berger, M. F. Loutre, and V. Dehant, "Pre-Quaternary Milankovitch frequencies." *Nature* 342:133 (1989).

A number of candidates for a biblical deluge have since appeared, such as the waterfall when the Mediterranean poured into the Black Sea, which 7,500 years ago was a freshwater lake. Few of the shoreline residents of the Black Sea knew anything more than that the shoreline kept moving inland, at speeds like those of a glacier surging. See Fagan (1999), pp. 87-88.

Charles Darwin, "Notes on the effects produced by the ancient glaciers of Caernarvonshire, and on the boulders transported by floating ice," *Philosophical Magazine,* vol. XXI (1842). For a chronology of discoveries about the ice ages, see Imbrie & Imbrie (1986), pp.195-202.
213 Joseph Adhémar history from Wallace S. Broecker and George H. Denton, "What drives glacial cycles?", *Scientific American* 262(1):48-56 (January 1990), at p. 49. Philander (1998), pp.173-174, has one of the best explanations of the Milankovitch "orbital" factors that I have seen. Figures adapted from Andre Berger, M. F. Loutre, H. Gallée, "Sensitivity

of the LLN climate model to the astronomical and CO_2 forcings over the last 200 ky," *Climate Dynamics* 14:615-629 (1998).

213 Peter U. Clark, Richard B. Alley, and David Pollard, "Northern Hemisphere ice-sheet influences on global climate change," *Science* 286: 1104-1111 (5 November 1999).

Alley (2000), p.98. Note that I am sidestepping the 100,000 year problem; the astronomy doesn't predict a major meltoff at such intervals, and there is much speculation about exotic and terrestrial causes. See, for example, Richard A. Muller, "Glacial cycles and orbital inclination," *Lawrence Berkeley Laboratory Report* LBL-35665 (1994), available at *http://www-physics.lbl.gov/www/astro/nemesis/LBL-35665.html*. Stochastic resonance is another possibility.

215 Discovery of Younger Dryas: Dorothy Peteet, "Sensitivity and rapidity of vegetational response to abrupt climate change," *Proceedings of the National Academy of Sciences (U.S.)* 97(4):1359-1361 (15 February 2000) at *http://www.pnas.org/cgi/content-/full/97/4/1359*

The temperature and precipitation records are replotted from the data of K. M. Cuffey and G. D. Clow, "Temperature, accumulation, and ice sheet elevation in Central Greenland through the last deglacial transition," *Journal of Geophysical Research* 102(C12):26383-96 (1997).

217 A well-written account of Hans Oeschger 's study of fluctuations seen in the ice cores can be found in Thomas Levenson, *Ice Times* (Harper & Row 1989), chapter 3. Thomas Stocker's obit of Oeschger can be found at *http://www.climate.unibe.ch-/oeschger/obituary.html.*

218 The Heinrich events demonstrate that the North Atlantic may have a third mode of ocean circulation. Today, and in the warm part of the D-O cycles, there is both far-north (above 60°N) and near-north sinking. During the cool-and-dry part of the D-O cycle, the far-north sinking shuts down. But during the Heinrich events, with so much fresh water being released as icebergs sail toward the Bay of Biscay, even the near-north sinking process shuts down, leaving warm water pooling in the southern oceans with no place to go. See Alley (2000), pp.153-155.

Kurt M. Cuffey and Shawn J. Marshall, "Substantial contribution to sea-level rise during the last interglacial from the Greenland ice sheet," *Nature* 404:591-594 (2000). The West Antarctic ice sheet contains enough water to raise sea level about 6 meters, as does Greenland; the East Antarctic ice sheet is huge in comparison, containing enough to raise sea level 60 meters. All other glaciers in the world combined contain about 0.5 meters worth. Both Greenland and West Antarctica have sunk enough under the weight of their ice that much of the ground beneath them is now below sea level, making them particularly vulnerable.

W. S. Broecker, D. Peteet, I. Hajdas, J. Lin, E. Clark, "Antiphasing between rainfall in Africa's Rift Valley and North America's Great Basin," *Quaternary Research* 50:12-20 (1998). "The beginning of the Bølling-Alleröd warm period was marked in Greenland by an abrupt rise in ^{18}O, an abrupt drop in dust rain, and an abrupt increase in atmospheric methane content. The surface waters in the Norwegian Sea underwent a simultaneous abrupt warming. At about this time, a major change in the pattern of global rainfall occurred. Lake Victoria (latitude 0°), which prior to this time was dry, was rejuvenated. The Red Sea, which prior to this time was hypersaline, freshened."

220 My sketch of the extent of grounded northern ice sheets is shown atop a satellite photo at modern sea levels with floating sea ice removed. The sketch is adapted from the one in Clark et al. (1999).

The plane where it's always noon

223 Time resolution on the scale of a year or two can be obtained from semi-fossil trees, e.g., Roig et al (2001) have a 1229-year-long stretch of tree-ring widths from the middle of the last ice age at 40°S that shows abrupt droughts with abrupt recoveries (tenfold changes in yearly accumulation), but they are floating in absolute time and such local records cannot yet be matched to events in the ice-core records.

Tropics affected: Konrad A. Hughen, Jonathan T. Overpeck, Scott J. Lehman, Michaele Kashgarian, John Southon, Larry C. Peterson, Richard Alley, & Daniel M.

Sigman, "Deglacial changes in ocean circulation from an extended radiocarbon calibration," *Nature* 391:65-68 (1 January 1998). The figure is adapted from their figure 2 (GISP2 accumulation record and Cariaco gray scale). And see the Cariaco Project at *http://paria.marine.usf.edu/cariaco/.*

J. P. Kennett and B. L. Ingram, "A 20,000 year record of ocean circulation and climate change from the Santa Barbara Basin," *Nature* 377: 510-517 (1995).

Climate change was synchronous (within a few decades) over a region of at least hemispheric extent when rewarming from the Younger Dryas: Jeffrey P. Severinghaus, Todd Sowers, Edward J. Brook, Richard B. Alley, & Michael L. Bender, "Timing of abrupt climate change at the end of the Younger Dryas interval from thermally fractionated gases in polar ice," *Nature* 391:141-146 (8 January 1998).

223 J. A. Eddy and Hans Oeschger, editors, *Global Changes in the Perspective of the Past* (Wiley 1993).

"Broecker, too, heard Oeschger that year": see Wallace S. Broecker, "Will our ride into the greenhouse future be a smooth one?" *GSA Today* 7(5):1-7 (May 1997). For more autobiography, see his "Converging paths leading to the role of oceans in climate change," *Annual Reviews of Energy and Environment* 25:1-19 (2000).

Droughts, see Fagan (1999), p.87.

M. E. Raymo, K. Ganley, S. Carter, D. W. Oppo, & J. McManus, "Millennial-scale climate instability during the early Pleistocene epoch," *Nature* 392:699-702 (16 April 1998).

224 "Relative stability" means avoiding the centuries-duration scale of D-O events of 5-15°C, though that leaves a lot of room for more minor fluctuations such as the 1,500-year cycle and the decade-scale rhythms that are now being studied. For some contrasts between the Younger Dryas and the most serious droughts since then at 8200, 5200, and 4200 years ago, see Fagan (1999) and Harvey Weiss, "Beyond the Younger Dryas: Collapse as adaptation to abrupt climate change in ancient West Asia and the Eastern Mediterranean," pp. 75-98 in *Confronting Natural Disaster: Engaging the Past to Understand the Future*, G. Bawden and R. Reycraft, editors (University of New Mexico Press 2000), at *http://www.yale.edu/nelc/weiss/byd.html.* **224**At *http://www.glaciology.gfy.ku.dk* is most of the North GRIP project information, and the earlier data is available at *http://arcss.colorado.edu/Gispgrip.*

P. E. Biscaye, F. E. Grousset, M. Revel, S. VanderGaast, G. A. Zielinski, A. Vaars, and G. Kukla, "Asian provenance of glacial dust (stage2) in the Greenland Ice Sheet Project 2 Ice Core, Summit, Greenland," *Journal of Geophysical Research* 102:765-781 (1997). Carbon dioxide is best studied in bubbles from Antarctic cores, where the ice is fewer potential contaminants than ice from Greenland; in general, see Alley (2000), p.103.

226 Pieter M. Grootes, M. Stuiver, J. W. C. White, S. Johnsen, and J. Jouzel, "Comparison of oxygen isotope records from the GISP2 and GRIP Greenland ice cores," *Nature* 366:552-554 (1993).

Willi Dansgaard, S. J. Johnsen, H. B. Clausen, D. Dahl-Jensen, N. S. Gundestrup, C. U. Hammer, C. S. Hvidberg, J. P. Steffensen, A. E. Sveinbjörnsdottir, J. Jouzel, G. Bond, "Evidence for general instability of past climate from a 250 kyr ice core," *Nature* 364:218-219 (1993).

Figure adapted from J.R. Petit, J. Jouzel, D. Raynaud, N.I. Barkov, J.-M. Barnola, I. Basile, M. Bender, J. Chappellaz, M. Davis, G. Delayque, M. Delmotte, V.M. Kotlyakov, M. Legrand, V.Y. Lipenkov, C. Lorius, L. Pépin, C. Ritz, E. Saltzman, and M. Stievenard, "Climate and atmospheric history of the past 420,000 years from the Vostok ice core, Antarctica," *Nature* 399:429-436 (1999).

228 Laurie Anderson, "Same time tomorrow" lyrics, *Bright Red*, track 14 (Time Warner 1994).

High above Oslo

230 Leonid Polyak, Margo H. Edwards, Bernard J. Coakley, Martin Jakobsson, "Ice shelves in the Pleistocene Arctic Ocean inferred from glaciogenic deep-sea bedforms," *Nature* 410:453-457 (22 March 2001). Bottom scouring at 1000 m depths; icebergs in the Arctic Ocean have at most 50-m draughts, whereas icebergs off Antarctica and Greenland reach depths of 550 m.

232 The reason I say "habitable world appears to be chilled" is that some areas of the South Pacific and Atlantic Oceans, and some (but not all) parts of Antarctica, may be exceptions. The extent of the Younger Dryas in higher southern latitudes may be somewhat spotty, as it is superimposed on the warming trend after the last glacial maximum which began in the south and spread north. See K. D. Bennett, S. G. Haberle, and S. H. Lumley, "The Last Glacial-Holocene transition in southern Chile," *Science* 290:325-328 (13 October 2000), *http://www.sciencemag.org/cgi/content/abstract/290/5490/325* and the perspective at pp.285-286.

Wallace S. Broecker, "Chaotic climate," *Scientific American* 273(5):62-69 (November 1995). A hundred Amazons is Broecker's figure; if the Amazon flow is instead taken as 0.19 Sv, and the southbound NADW off Newfoundland as 13 Sv, then it is more like 70 Amazons. But deep water production by convection may be less, depending on how much NADW is Arctic in origin and how much is simply recirculated Antarctic bottom water (extremely dense water, formed as brine under the sea ice around polynas offshore of Antarctica and sliding down the continental shelf into the depths without much mixing, creates a giant pool of dense water extending all the way up the bottom of the Atlantic to about 60°N). Greenland Sea production is said to be 5.6 Sv, about equally divided between the sills to the east and west of Iceland. Labrador Sea production is difficult to estimate because of the local gyre and the frequent annual failures.

233 Robert Bindschadler, "Future of the West Antarctic Ice Sheet," *Science* 282:428-429 (16 October 1998).

G. H. Denton, C. H. Hendy, "Younger Dryas Age Advance of Franz Josef Glacier in the Southern Alps of New Zealand," *Science* 264:1434-1437 (3 June 1994; follow up at 271:668-669, 2 February 1996).

Judith Lean and David Rind, "Earth's Response to a Variable Sun," *Science* 292(5515) 234-236 (13 April 2001).

Peter V. Foukal, "The variable sun," *Scientific American* 262(2):34-41 (February 1990).

U. Neff, S. J. Burns, A. Mangini, M. Mudelsee, D. Fleitmann & A. Matter, "Strong coherence between solar variability and the monsoon in Oman between 9 and 6 kyr ago," *Nature* 411:290-293 (2001).

J. D. Haigh, "The impact of solar variability on climate," *Science* 272:981-984 (1996).

Reductions in solar output might, of course, be one of the things that set the stage for an abrupt cooling, as might the North Atlantic Oscillation and other atmospheric cycles like El Niño. Causes usually don't come one at a time, but combine in various ways.

For climate variability more generally, start at the CLIVAR web pages at *http://www.clivar.org/start.htm*

236 Alley (2000), p.3.

Out over the sinking Gulf Stream

239 Henry M. Stommel, "Thermohaline convection with two stable regimes of flow," *Tellus* 13:224-230 (1961) had the concept of freshening of surface waters interfering with the thermohaline sinking. The conceptual map is from Stommel (1957), see *http://www.aoml.noaa.gov/hrd/Landsea/climtrend/.*

240 Wallace S. Broecker, "The great ocean conveyor," *Oceanography* 4:79-89 (1991).

Wallace S. Broecker, "Abrupt climate change: causal constraints provided by the paleoclimate record," *Earth-Science Reviews* 51:137-154 (August 2000).

D. J. Webb and N. Suginohara, "Oceanography: Vertical mixing in the ocean," *Nature* 409:37 (4 January 2001).

Stefan Rahmstorf, "Climate, Abrupt Change," chapter in *Encyclopedia of Ocean Sciences*, edited by J. Steele, S. Thorpe and K. Turekian (Academic Press, London, 2001).

242 Alley (2000), p.4; Broecker (1997), p.1; Stefan Rahmstorf (2001) , essay at *http://www.pik-potsdam.de/~stefan/essay.html.*

Jan Mayen Island

243 Walter Munk, Laurence Armi, Kenneth Fischer, F. Zachariasen, "Spirals on the sea," *Proceedings of the Royal Society: Mathematical, Physical & Engineering Sciences* 456:1217-1280

(1997). First seen in the Apollo missions of the 1960s, spirals are broadly distributed over the world's oceans, 10-25 km in size and overwhelmingly cyclonic.

Whirlpools associated with salt play an interesting role in Norse legends. See chapter 6 of Giorgio de Santillana and Hertha von Dechend, *Hamlet's Mill: An Essay Investigating the Origins of Human Knowledge and its Transmission Through Myth* (Godine 1969).

244 Alexander Sy, Monika Rhein, John R. N. Lazier, Klaus Peter Koltermann, Jens Meincke, Alfred Putzka, and Manfred Bersch, "Surprisingly rapid spreading of newly formed intermediate waters across the North Atlantic Ocean," *Nature* 386:675-679 (17 April 1997).

Michael S. McCartney, Ruth G. Curry, Hugo F. Bezdek, "North Atlantic transformation pipeline chills and redistributes subtropical water," *Oceanus* 39(2):19-23 (Fall/winter 1996). This article elaborates on the multi-year aspects of the formation of salt sinking in the sub-polar gyre.

The Greenland Sea

248 Switching from Greenland to Labrador domination: R. Dickson, J. Lazier, J. Meincke, P. Rhines, J. Swift, "Long-term coordinated changes in the convective activity of the North Atlantic," *Progress in Oceanography* 38:241-295 (1996).

G. C. Bond, R. Lotti, "Iceberg discharges into the North Atlantic on millennial time scales during the last glaciation," *Science* 267:1005 (17 February 1995).

Wallace S. Broecker, "Massive iceberg discharges as triggers for global climate change," *Nature* 372:421-424 (1994).

249 Date for closure of "Old Panama Canal": Neil D. Opdyke, "Mammalian migration and climate over the last seven million years," chapter 8 in *Paleoclimate and Evolution, with Emphasis on Human Origins*, edited by Elisabeth S. Vrba, George H. Denton, Timothy C. Partridge, Lloyd H. Burckle (Yale University Press 1995), pp.109-114 at p.112.

250 C. H. Haug, R. Tiedemann, "Effect of the formation of the Isthmus of Panama on Atlantic Ocean thermohaline circulation," *Nature* 393(6686):673-676 (18 June 1998).

There is a similar theory for how African climate was affected by New Guinea moving northward to close off the easy circulation between Pacific and Indian Oceans: Mark A. Cane, Peter Molnar, "Closing of the Indonesian seaway as a precursor to east African aridification around 3–4 million years ago," *Nature* 411:157-162 (10 May 2001).

252 Photograph of Greenland mountain peaks exposed by the thinning continental ice sheet: Rock protruding through an ice sheet is called a nunatak. The low light angle is from the midnight sun, as the picture was taken looking south on an eastbound flight in the summer in the early morning hours.

Greenland fjords

254 Peter Schlosser, Gerhard Bönisch, Monika Rhein, Reinhold Bayer, "Reduction of deepwater formation in the Greenland Sea during the 1980s: Evidence from tracer data," *Science* 251:1054-1056 (1 March 1991).

Gerhard Bönisch, Johan Blindheim, John L. Bullister, Peter Schlosser, and Douglas W. R. Wallace, "Long term trends of temperature, salinity, density, and transient tracers in the Central Greenland Sea," *Journal of Geophysical Research* 102:18553-18571 (1997), see *http://www.ldeo.columbia.edu/~noblegas/gerhard/GIN/tspgs/tsp65.html.*

Michael S. McCartney, "North Atlantic Oscillation," *Oceanus* 39(2):13 (Fall/winter 1996). The regional atmospheric circulation over the North Atlantic normally features a high over the Azores and a low near Greenland and Iceland - the westerlies are intense but the cold air from Canada is warmed before reaching Europe. When the low shifts as far south as Newfoundland, a high develops over northern Greenland; this brings cold arctic air west from northern Europe to be warmed by the Norwegian Current and thus warm Greenland and North America rather than Europe. The Labrador Sea is much less stormy and this likely affects salt sinking.

The NAO is best understood as a ring of pressure anomalies extending around the globe, not just a North Atlantic phenomenon (something similar happens in the Southern Hemisphere at similar latitudes). See Dennis L. Hartmann, John M. Wallace, Varavut

Limpasuvan, David W. J. Thompson, and James R. Holton, "Can ozone depletion and global warming interact to produce rapid climate change?" *Proceedings of the National Academy of Sciences (U.S.)* 97:1412-1417 (15 February 2000) at *http://www.pnas.org/cgi/content/full/97/4/1319.*

254 The relationships between the NAO and deep water production are discussed by R. Dickson, "Observations of DecCen climate variability in convection and water mass formation in the northern hemisphere," in the CLIVAR Villefranche workshop summary at *http://www.dkrz.de/clivar/villesum.html.* More generally, see the Climate Research Committee, National Research Council, *Natural Climate Variability on Decade-to-Century Time Scales* (National Academy Press 1995). Excerpts can be found at *http://www.nap.edu/bookstore/isbn/0309054494.html.*

255 Sinking to the very bottom may be important for the return leg of the trip, south past the coast of Newfoundland where it must stay beneath the Gulf Stream. There is some speculation that it sometimes doesn't, and is swept eastward, illustrating a mechanism for creating blobs that recirculate. It is known that semi-salty, or anomalously warm or cool, blobs even travel from the subtropics, up and around the Greenland Seas and back around to the Labrador Sea, suggesting another failure mode for late winter downwelling; see Raymond W. Schmitt, "If rain falls on the ocean, does it make a sound?" *Oceanus* 39(2):4-8 (Fall/winter 1996).

256 Thomas F. Stocker, Olivier Marchal, "Abrupt climate change in the computer: Is it real?" *Proceedings of the National Academy of Sciences (U.S.)* 97:1362-1365 (15 February 2000) at *http://www.climate.unibe.ch/~stocker/.*

Andrey Ganopolski and Stefan Rahmstorf, "Simulation of rapid glacial climate changes in a coupled climate model," *Nature* 409:153-158 (11 January 2001) at *http://www.pik-potsdam.de/~stefan/.*

Till Kuhlbrodt, Sven Titz, Ulrike Feudel, Stefan Rahmstorf, "A simple model of seasonal open ocean convection. Part II: Labrador Sea stability and stochastic forcing," *Ocean Dynamics* (in press, 2001). They conclude that the Labrador Sea is probably in a bistable regime and investigate the effect of stochastic forcing.

Bogi Hansen, William R. Turrell, Svein Østerhus, "Decreasing overflow from the Nordic seas into the Atlantic Ocean through the Faroe Bank channel since 1950," *Nature* 411:927-930 (21 June 2001).

Thomas Levenson, *Ice Times* (Harper & Row 1989), p.49.

Atop Greenland

258 For the hinge from which icebergs break off, see E. J. Rignot, S. P. Gogineni, W. B. Krabill, S. Ekholm, "North and northeast Greenland ice discharge from satellite radar interferometry," *Science* 276:934-937 (9 May 1997). *http://www.sciencemag.org/cgi/content/full/276/5314/934*

Sometimes glaciers surge a mile a month (the Bering glacier east of Cordova, Alaska - North America's longest glacier - is currently surging), but that's because of warming, not cooling. Another one of the ice paradoxes is that glaciers can advance for either reason. Glaciers, too, have modes of operation: once melting starts, a mountain of ice tends to collapse, spreading sideways. Because meltwater runs off beneath the glacier (and can form there from all the trapped heat from the earth), it smoothes the undersurface and greases the skids. The glacial snout may advance so quickly that the ocean's chipping away at it cannot keep up. And so the snout of the glacier pushes further into the sea over a course of months. Eventually the tides may break it loose.

William S. Reeburgh, D. L. Nebert, "The birth and death of Russell Lake," *Alaska Science Forum* 832 (3 August 1987) at *http://www.gi.alaska.edu/ScienceForum/ASF8/832.html*

259 R. B. Alley, P. A. Mayewski, T. Sowers, K. C. Taylor, and P. U. Clark, "Holocene climatic instability: A prominent, widespread event 8200 yr ago." *Geology* 25:483-486 (1997).

D. C. Barber, A. Dyke, C. Hillaire-Marcel, A. E. Jennings, J. T. Andrews, M. W. Kerwin, G. Bilodeau, R. Mcneely, J. Southon, M. D. Morehead & J.-M. Gagnon, "Forcing of the cold event of 8,200 years ago by catastrophic drainage of Laurentide lakes," *Nature* 400:344-348 (1999).

260 Richard A. Kerr, "Warming's unpleasant surprise: Shivering in the greenhouse?" *Science* 281:156-158 (news article, 10 July 1998).

Thule

262 An even longer record comes from the high tropics, going back about 500,000 years and showing many abrupt temperature changes. L. G. Thompson, T. Yao, M. E. Davis, K. A. Henderson, E. Mosley-Thompson, P.-N. Lin, J. Beer, H.-A. Synal, J. Cole-Dai, J. F. Bolzan, "Tropical climate instability: The last glacial cycle from a Qinghai-Tibetan ice core," *Science* 276:1821-1825 (20 June 1997) at *http://www.sciencemag.org/cgi/content/full/276/5320/1821.*

Mark Maslin, "Sultry last interglacial gets a sudden chill," *Earth in Space* 9(7):12-14 (March 1997) at *http://www.agu.org/sci_soc/eismaslin.html.* One of the current uncertainties about the mid-Eemian cooling event at 122,000 years is whether it involved the complete shutdown of the North Atlantic Deep Water production; various ocean-floor cores suggest it didn't.

L. D. Keigwin, W. B. Curry, S. J. Lehman, S. Johnsen, "The role of the deep ocean in North Atlantic climate change between 70 and 130 kyr ago," *Nature* 371:323-326 (22 September 1994).

J. F. McManus, G. C. Bond, W. S. Broecker, S. Johnsen, L. Labeyrie, S. Higgins, "High-resolution climate records from the North Atlantic during the last interglacial," *Nature* 371:326-329 (22 September 1994).

Robert R. Dickson, "The local, regional, and global significance of exchanges through the Denmark Strait and Irminger Sea," in *Natural Climate Variability of Decade-to-Century Time Scales* (National Academy Press 1995), pp. 305-317.

R. R. Dickson, J. Meincke, S-A. Malmberg, and A. J. Lee, "The 'great salinity anomaly' in the northern North Atlantic 1968-1982," *Progress in Oceanography* 20:103-151 (1988). Note that the Great Salinity Anomaly, equivalent to an extra 2,000 km^3 of fresh water, is not usually treated as a fjord flood but as a variant on a semi-salty current out of the Arctic Sea. This volume is about 500 times the freshwater flux from Alaska's Russell Fjord over a five-month period - but then the reservoir capacity of the east Greenland fjord system at 70-77°N is also extraordinary and ice dams can span multiple years. The travel time from the Labrador Sea to Bermuda is about six years.

264 Hard to restart: it's an example of hysteresis. This is one of those asymmetric situations where it is easy to get in but hard to get out, something like a one-way street where you can't turn around but are forced to go all the way around the block instead. See the diagram in Stocker & Marchal (2000).

265 The 1,500-year solar cycle: Gerard Bond, Bernd Kromer, Juerg Beer, Raimund Muscheler, Michael N. Evans, William Showers, Sharon Hoffmann, Rusty Lotti-Bond, Irka Hajdas, and Georges Bonani, "Persistent Solar Influence on North Atlantic Climate During the Holocene," *Science* 294(5549)2130-2136 (7 December 2001).

Note also that desert dust inhibits precipitation by a mechanism that increases albedo: Daniel Rosenfeld, Yinon Rudich,, and Ronen Lahav, "Desert dust suppressing precipitation: A possible desertification feedback loop," *Proceedings of the National Academy of Sciences (U.S.)* 98(11):5975-5980 (22 May 2001).

Windy-and-dusty increasing ocean productivity, reviewed by Sallie W. Chisholm, Paul G. Falkowski, and John J. Cullen, "Dis-crediting ocean fertilization," *Science* 294:309-310 (12 October 2001), at *http://www.sciencemag.org/cgi/content/summary/294/5541/309.*

268 Ray T. Pierrehumbert, "Climate change and the tropical Pacific: The sleeping dragon wakes," *Proceedings of the National Academy of Sciences (U.S.)* 97:1355-1358 (10 February 2000) at *http://www.pnas.org/cgi/content/full/97/4/1355*

Somerset Island

269 For more on why open ocean occurs occasionally in Arctic summers, sometimes even at the pole itself, see *http://psc.apl.washington.edu/northpole/NPOpenWater.html.* There is an enormous heat flux through them, as the difference between surface and air temperature is 30°C. They soon refreeze (unless they are maintained open by offshore winds, as in Antarctica) and produce a layer of brine.

News article by John Noble Wilford, "Ages-old polar icecap is melting, scientists find," *New York Times* (August 19, 2000), at *http://www.nytimes.com/library/national /science/081900sci-climate-pole.html.* Actually, none of the Arctic sea ice is much over three years old, and that at 90°N is usually only a year old. The more common confusion is between an ice sheet, which can be 3,000 meters thick, and floating sea ice (in the Arctic, seldom more than 3 meters thick). The Antarctic is a continent surrounded by ocean; the Arctic is a nearly landlocked sea surrounded by continents which empty fresh water into it via north-flowing rivers; using the same word, "icecap," for both is misleading, however correct.

270 D. A. Rothrock, Y. Yu, G. A. Maykut, "Thinning of sea-ice cover," *Geophysical Research Letters* 26:3469-3472 (1 December 1999).

Ola M. Johannessen, Elena V. Shalina, Martin W. Miles, "Satellite evidence for an Arctic sea ice cover in transformation," *Science* 286:1937-1939 (3 December 1999).

Konstantin Y. Vinnikov, Alan Robock, Ronald J. Stouffer, John E. Walsh, Claire L. Parkinson, Donald J. Cavalieri, John F. B. Mitchell, Donald Garrett, Victor F. Zakharov, "Global warming and Northern Hemisphere sea ice extent," *Science* 286:1934-1937 (3 December 1999). And see the follow up at *Science* 288:927 (2000).

The special pleading works much better with real juries, from which knowledgeable jurors have been systematically excluded, than it does with the general public. Anyone who knows more about the science or technology than the lawyers is usually rejected during jury selection. And the judges allow it.

Instrumental record: J. Fred Singer, "Global warming: An insignificant trend?" *Science* 292:1063 (2001). And see Donald Kennedy's pointed reply which follows it.

271 Abrupt change from wet to dry in the Sahara (at least, as measured by offshore dust) as the axial tilt gradually changes: Peter B. deMenocal, J. Ortiz, T. Guilderson, J. Adkins, M. Sarnthein, L. Baker, and M. Yarusinsky, "Abrupt onset and termination of the African Humid Period: Rapid climate responses to gradual insolation forcing," *Quaternary Science Review* 19: 347-361 (2000).

Quote from Alley (2000), p. 83.

272 Though I cannot find any literature on equatorial warming triggering reorganization for the D-O events, there are reports, for the glacial-interglacial transition, that Pacific sea surface temperatures warmed 3,000 years before changes in ice volumes. See David W. Lea, Dorothy K. Pak, and Howard J. Spero, "Climate impact of Late Quaternary equatorial Pacific sea surface," *Science* 289:1719 (8 September 2000) at *http://www.sciencemag.org/cgi- /content/full/289/5485/1719.*

Quote from Broecker (1997), p. 5.

274 Philander (1999), p.3.

Crossing the North American coast

278 Apropos the comparison of Canada to Europe, Conway Leovy reminds me that Europe does not have an equivalent of the rain shadow from the Rockies.

281 Edward O. Wilson, *Consilience: Unity of Knowledge* (Knopf 1998), p.13.

282 Weiner (1994), p.286.

Yellowknife, Northwest Territories

283 Canadian warming: K. E. Taylor, J. E. Penner, "Response of the climate system to atmospheric aerosols and greenhouse gases," *Nature* 369:734-737 (1994).

The idea of a Precambrian glaciation covering even the tropics goes back to the 1960s but it was dismissed, despite the evidence of glacial till in suspicious places, overlain by tropical sediments. Essentially, should the sun's output reaching the earth decline a few percent (Sun itself, dust clouds, whatever), even the tropics freeze and the albedo keeps it that way. The idea was revived in the 1990s because of a mechanism (volcanic CO_2 becoming 10% of air) that would melt it back within a few million years, after which there is a Hothouse Earth for a few million years until the thirty-fold-greater CO_2 is removed from the atmosphere. At least two such episodes are thought to have occurred between 750 and 550 million years ago.

Paul F. Hoffman, Alan J. Kaufman, Galen P. Halverson, and Daniel P. Schrag, "A neoproterozoic snowball earth," *Science* 281(5381):1342-1346 (28 August 1998).

Paul F. Hoffman, Daniel P. Schrag, "Snowball Earth," *Scientific American*, pp. 68-75 (January 2000) at *http://www.sciam.com/2000/0100issue/0100hoffman.html*

284 Weathering feedback, see Alley (2000), p.86.

285 Robert J. Charlson, James E. Lovelock, Meinrat O. Andreae, Stephen G. Warren, "Oceanic phytoplankton, atmospheric sulphur, cloud albedo, and climate." *Nature* 326:655-661 (1987).

287 Noise in threshold systems, and the first passage time, was the topic of my very first scientific paper: W. H. Calvin, C. F. Stevens, "A Markoff process model for neuron behavior in the interspike interval," *Proceedings of the Annual Conference on Engineering in Medicine and Biology* 7:118 (1965).

Bumpy border crossing

288 Lewis Thomas, *Late Night Thoughts on Listening to Mahler's Ninth Symphony*, (Viking 1983), p. 15.

289 Robert Burns, *The Shape and Form of Puget Sound* (University of Washington Press 1985), p. 43.

Michael Parfit, "Before Noah, there were the Lake Missoula Floods," *Smithsonian* (April 1995).

John Eliot Allen and Marjorie Burns, *Cataclysms on the Columbia* (Portland: Timber Press 1986).

290 See Antonio Damasio, *The Feeling of What Happens* (1999). "Consciousness and its revelations allow us to create a better life... but the price we pay for that better life is high. It is not just the price of risk and danger and pain. It is the price of *knowing* risk, danger, and pain. Worse even: it is the price of knowing what pleasure is and *knowing* when it is missing or unattainable." [p.316]

Afterthoughts

297 William H. Calvin, "The great climate flip-flop," *The Atlantic Monthly* 281(1):47-64 (January 1998) at *http://faculty.washington.edu/wcalvin/1990s/1998AtlanticClimate.htm.*

The underreporting of abrupt climate change persisted from 1985 to 1998, even in the face of substantial recognition of some of the major players. For example, Wallace S. Broecker - easily the most vigorous of the geoscientists in trying to alert the scientific community, and author of several *Scientific American* articles - was awarded the U.S. National Medal of Science by President Clinton in 1996 for "contributions to understanding chemical changes in the ocean and atmosphere." (*http://imager.ldeo.columbia.edu/geol_sci/html/wallace_broecker.html*). The Danish ice-core expert Willi Dansgaard and the British oceanographer Nicholas Shackleton received the Crafoord Prize from the Swedish Academy in 1995 (see *http://www.kva.se/prizes.html*). Dansgaard, the Swiss climatologist Hans Oeschger, and the French climatologist Claude Lorius received the $150,000 Tyler Prize in 1996.

Yet, despite all this recognition and all those news stories in *Science* and *Nature*, the *bistable* climate story itself (sudden warmings, flipping to sudden coolings) was seldom reported in the popular press. It was conflated with other rapid climate changes (volcanoes, ice shelves breaking up) lacking bistable states, or it was simply lost in a greenhouse story. It would be interesting for some student of the media to sort out the underlying reasons why such a major story was ignored.

An exception was Ross Gelbspan, *The Heat Is On* (Addison-Wesley 1997), who treats the abrupt cooling story at pp. 28-31. For a modern example of good reporting, see Fred Pearce, "Violent future," *New Scientist* 2300:4-5 (21 July 2001).

"Anthropogenic warming" is a classic example of how *not* to frame an issue. It hurts no matter what caused it; assigning fault doesn't resolve anything in this case. And just because something is "natural" doesn't mean it is out of our control. (Floods? Smallpox?) Many of the people whose pockets might be hurt by expenses for combating warming are quite happy to see the issues dumped solely into the laps of the scientists, perhaps because they know that scientific studies always raise more questions than they answer, and so the issue of whether science has spoken on the issue can be indefinitely postponed. Public policy has some similarities to medicine; the physician who waits until dead certain of a diagnosis before starting treatment may often wind up with a dead patient. Often, one cannot wait for scientific certainty.

Glossary

309 Peter Schuster, "Molecular insights into life and evolution," the 1998 Schrödinger lecture, at Trinity College Dublin, *http://www.tcd.ie/Physics/Schrodinger/Lectures/lecture4.html.*

Subject Index

The book's full text is on web pages at *WilliamCalvin.com* and the search engines have indexed every word, far better than I can do it. In my abbreviated index below, you generally will not find most high-frequency words like "hunting" or "Darwin."

Glossary entries are in **bold**. If you can't find what you want below, just use my search form at *http://WilliamCalvin.com/search.*